Information
Theoretic
Indices for
Characterization
of Chemical
Structures

CHEMOMETRICS SERIES

Series Editor: **Dr. D. Bawden**
Pfizer Central Research, Sandwich, Kent, England

Information Theoretic Indices for Characterization of Chemical Structures

Professor Danail Bonchev
Higher School of Chemical Technology, Burgas, Bulgaria

RESEARCH STUDIES PRESS
A DIVISION OF JOHN WILEY & SONS LTD
Chichester · New York · Brisbane · Toronto · Singapore

RESEARCH STUDIES PRESS

Editorial Office:
58B Station Road, Letchworth, Herts. SG6 3BE, England

Copyright © 1983, by John Wiley & Sons Ltd.

Library of Congress Cataloging in Publication Data:

Bonchev, Danail.
    Information theoretic indices for characterisation of
    chemical structures.
    (Chemometrics research studies series; no. 5)
    Includes index.
    1. Chemical structure.   2. Information theory in
    chemistry.   I. Title.   II. Series.
    QD471.B76 1983   541.2'2   82-16696
    ISBN 0 471 90087 7

British Library Cataloguing in Publication Data:

Bonchev, Danail.
    Information theoretic indices for characterisation of
    chemical structures—(Chemometrics research studies
    series; 5)
    1. Chemistry—Information services
    2. Chemical literature
    I. Title
    540'.7   Q08.3

    ISBN 0 471 90087 7

Printed in Great Britain

Editorial Preface

The concepts of information theory have been steadily
infiltrating the physical and biological sciences for
several decades now. Although originally conceived as a
means of modelling communications channels, information
theory has held out the prospect of the introduction of
quantitative measures of "order", "complexity", and "information content" into scientific thought. Chemistry,
with its hierarchy of nuclei, atoms, functionabilities,
molecules, and polymers, and its highly varied, though
ordered, patterns of bonding and structure, is a particularly attractive area for the application of information
theory.

Professor Bonchev here presents such an application,
the derivation of numerical indices based on information
theoretic formulae to represent the topological structure
of atoms and molecules. Indices based on a graph theoretical analysis of molecular structure have a relatively
long history, whereas the information theoretic indices
are of more recent introduction. The number of such indices
is quite large, to the uninitiated confusingly so. Professor Bonchev, in thoroughly reviewing and comparing these
indices, sets them clearly into content and, by including
the results of his more recent work, shows the directions
in which the field is moving.

The representation of atomic and molecular structure by numeric indices has great attraction, particularly when these indices may be related to concepts such as "complexity". The practical applications, particularly in the quantitative relating of structure to atomic or molecular properties, are beginning to become established, and are likely to increase in importance in the future.

Finally, it may be, as Profesor Bonchev notes, that this area of study may be a forerunner of an increasing tendency towards the unification of presently diverse scientific disciplines, through the common language of cybernetics and information theory. This is indeed an intriguing prospect.

David Bawden

Preface

Contemporary science is characterized by a high degree of
differentiation. The enormous number of new scientific
disciplines, which has arisen in the 20th century, hinders
the contacts between the different branches of science
threatening it with the ghost of Babel. Fortunately, in-
tegrating tendencies have also been in progress during the
last few decades including here many general sciences
such as cybernetics, theory of systems, artificial intelligence, information theory, etc.

Information theory can be regarded as a universal language for describing systems, allowing useful analogies
or common laws between systems of different nature to be
inferred. In this way the laws in a certain scientific
area, properly translated into information language, could
be projected into other, not well developed scientific
areas. The information approach however is not simply a
translation from less general languages to a universal one.
It provides additional insight into systems and phenomena,
allowing new results to be obtained.

The applicability of information theory to chemistry is
mainly based on the possibilites offered by it for the
quantitative analysis of various aspects of chemical
structures. Related to this, the present book is concer-
ned with the so-called theoretic information indices which

are structural indices expressing in a quantitative form
the degree of complexity of molecules and atoms. These indices are a powerful tool for discrimination between similar structures which make them fairly suitable for various
classificational aims, as well as for computer processing
and retrieval of chemical information. They may also be
helpful when a choice between competing approximations of
hypotheses has to be made. The most promising field of
application, however, is the correlation between structure
and properties (including biological activity) of chemical
systems, due to the potential possibility of translating
any structural features into a numerical descriptor, a
theoretic information index. Finally, new results and
generalizations may be expected on the basis of the extremal values of the information function.

This monograph is a survey of the information theoretic
indices introduced during the last three decades to
characterize molecules and atoms. To a large extent it
reports original results obtained in the theoretical chemistry group in Burgas, Bulgaria, where systematic studies
on the chemical applications of Information Theory have
been carried-out ever since 1968. An outline of the development of the notion of information is given in the Introduction. Other chemical applications, which are not explored further, are also reviewed there. The Second Chapter
deals briefly with the basic equations of Information Theory and the ways they are applied to structural problems
in Chemistry. The various atomic and molecular information-
theoretic indices are discussed in detail in Chapters III
to V, while their specific applications in Chapter VI. The
book has many examples and illustrations. The mathematical
formalism is reduced to a reasonable degree. This makes the
information theoretic approach accessible to every chemist,
in accord with the purpose of the book to stimulate the
interest of the chemical community towards this new branch

of mathematical chemistry.

The author would like to acknowledge his great indebtedness to Prof. Balaban (Bucharest), Prof. Trinajstić (Zagreb), Prof. Randić (Ames), Prof. Zhdanov (Rostov na Don), Prof. Dosmorov (Omsk), Dr. Roy (Calcutta) and Dr. Basak (Duluth, Minnesota), for offering him the opportunity to read their papers prior to publication, to Prof. Polansky (Muelheim) and Prof. Rouvray (London) for some useful criticism, to Dr. Nikolov (Sofia) for correcting the English of the manuscript, and to his co-workers Dr. Ovanes Mekenayn, Dr. Verginia Kamenska, and Dr. Brezitsa Rousseva for contributing to the development of some information-theoretic indices.

The author is particularly indebted to Dr. David Bawden of Pfizer Central Research, the Editor of this Series, who stimulated the writing of this monograph and encouraged its preparation.

Danail Bonchev

Table of Contents

CHAPTER 1
The Notion of Information and its Applications

The notion of information appears as one of the most fundamental notions in the 20-th century science, a notion of no less importance than that of matter and energy. This assertion follows from the very definition of information. According to Norbert Wiener (1948) "information is neither matter, nor energy". Ashby (1956) treats information as "a measure of the variety in a given system". Following Glushkov (1964) *"information is a measure of the nonhomogeneity in the distribution of matter or energy in space and time."* On this basis it becomes more and more evident that besides their substance and energy essence, the objects and phenomena in nature and technology also have an information character. Moreover, the centre of scientific research is expected by some prognoses to move towards the information nature of the processes as the major field of study in 21st century. These prognoses are essentially based on the possibility of systems and processes to be controlled by information, which is the major function of information in cybernetics. As a perspective these ideas could lead to a technology in which every atom or molecule is controlled by information, a possibility that is realized in living nature.

The year 1948 is usually considered as the birth date of Information Theory, since this was the year in which Claude

Shannon published his fundamental work (Shannon and Weaver, 1948). The concept of information as a quantity related to entropy is, however, much older. Boltzmann (1894) stated that every piece of information obtained for a physical system is related to the decrease in the number of its possible states, therefore the increase of entropy means "loss of information". Szillard (1929) developed this idea for the general case of information in physics. Later, Brillouin (1956, 1964) generalised the concept of entropy and information in his negentropy principle of information. He specified information as a negative entropy (negentropy) and extended the definition of the Second Law of Thermodynamics so as to encompass the information as well. The possible interplay between Information Theory and Thermodynamics and in particular between entropy and information, is a subject of permanent interest. (A list of selected references to this field is given elsewhere (Bonchev and Lickomannov, 1977). New horizons have been revealed for statistical thermodynamics on the basis of the Jaynes (1957 a,b) maximum entropy hypothesis (or Jaynes' principle). From the latest development of the problem the works of Kobozev (1971) on thermodynamics of thinking are of considerable interest where the concept of the anti-entropy character of thinking is proposed.

Arising as "a special theory of communications", Information Theory soon exceeded its initial limits and found application in numerous scientific and technical areas: physics, chemistry, biology, medicine, linguistics, psychology, aesthetics et al. (A detailed survey is given by Mathai and Rathie, 1975). The role of information has been recognized first in biology. Some important problems of conservation, processing and transmission of information in living beings have been solved, such as coding of genetic information (Quastler, Editor, 1953; Hasegawa and Yano,

1975; Seybold, 1976; Gatlin, 1972), estimation of the
possibility of spontaneous self-generation of life on
Earth (Rashevsky, 1960: Eigen and Winkler, 1975; Eigen and
Schuster, 1979), Formulation of the fundamental laws of
biological thermodynamics (Trincher, 1964), analysis of
the problems of bioenergetics (Bykhovskii, 1968), etc.
The information content of systems has been used as a
quantitative criterion of evolution (Ursul, 1966; Bonchev,
1970). The information character of food-consuming processes has been pointed out to dominate over their substance
and energy nature (Schroedinger, 1944; Brekhman, 1976).
Works on the information theory of disease diagnostics and
therapy are in progress.

The interest of the chemical community in Information
Theory has been permanently growing. Levine, Bernstein et
al. developed an information-theoretic approach to molecular dynamics which describes the behaviour of interacting
molecular systems far from equilibrium, in vibrationally
and rotationally excited states (Bernstein and Levine,
1972; Ben-Shaul et al., 1972; Levine and Bernstein, 1974,
1976). A quantitative measure of the information content,
or entropy deficiency, of different classes of experiments
has been introduced, e.g. molecular beam scattering, IR
chemoluminescence etc. This approach found various applications including the determination of branching ratios for
alternative reaction paths, the study on the operational
characteristics of lasers, etc. Equivalence has been
demonstrated to exist between the maximum entropy approach
to molecular collisions and their description by means of
the dynamical equations of motion (Alhassid and Levine,
1977).

4

An important contribution to quantum chemistry has been
made by Daudel et al. who developed the so-called loge
theory (Aslangul et al., 1972). A loge is a part of the
atomic or molecular space where there is high probability
of finding a certain number of electrons with a certain
organization of their spin. The best partition of the
space in loges is found by minimization of the Shannon
function which reduces to minimum the uncertainty about
the atom or molecule. The loge theory is of importance
for the quantum mechanical elucidation of the main chemical
idea, and first of all of the additivity of molecular
properties. This theory has also been applied to improve
approximate wave functions. The last idea has also been
explored by Maroulis et al. (1981) who applied Information
Theory to optimize the basis set wave functions in ab
initio calculations. Jaynes' principle has been applied by
Larson (1973) to construct reference density matrices for
studies of electron correlation. Sears et al. (1980) have
established precise connection between the quantum mechanical kinetic energy and the information on the electron
density distributions. On this basis the quantum mechanical variation principle has been defined as a principle of
minimal information. The Schrödinger equation has been
qualified as a statistical procedure for determining the
missing locality information inherent in quantum systems.

Information Theory is extensively applied to analytical
chemistry whose main goal is to acquire information on
chemical systems. Obtaining information is said to be the
object of chemical and instrumental analysis. Hence Information Theory was found very useful in the estimation of
the efficiency of analytical methods, minimization of
errors and analytical time, achievement of higher selectivity, or more generally, in optimizing the processes
of obtaining information in analytical chemistry (Ecksch-

lager, Stepanek, 1979).

The concept of combinatorial entropy of formation, which contributes to the stability of molecules, has been introduced by Gordon and Scantlebury (1964). It has been generalized by Gordon and Temple (1970) by presenting this entropy as consisting of a topological part and a metrical part. The latter is defined in terms of molecular size, shape, mobility, and location in the three-dimensional space while the former is defined in terms of graph theory, enumerating the possible molecular "graph-like" states.

Other fields of chemistry have also been subjects of information treatment. The sensitivity and catalytic activity of catalysts have been connected with their information content by Kobozev et al. (Kobozev et al., 1971 a,b; Beskov and Solyakina, 1975). The optimum information conditions for characterization and prediction of catalyst properties have also been formulated. An information analysis has been made of the surface treatment with different chemical reagents (Rackow, 1967 a,b; 1969). The formation and growth of crystals has been treated as an information process (Petrov, 1970 a). This approach has a large application to geochemistry for the characterization of frequency distributions (Vistelius, 1964), the determination of complexity (Petrov. 1970 b), and the general classifications of geochemical systems (Petrov, 1971) etc.

In this book the attention is centred on the information theoretic description of molecules and atoms. Various information indices are presented that characterize molecular chemical composition, topology, symmetry, etc., as well as the electronic distributions in atoms and molecules.

CHAPTER 2
Some Remarks on the
Information-Theoretic Approach.
The Concept of Structural Information

Information Theory offers quantitative methods for the
study of receiving, conserving, processing, transmitting
and practical use of information. The quantitative meas-
urement of information takes an important place in the information approach. The definition of the *quantity of information* requires, however, a full rejection of the wi-
despread but obscure qualitative ideas of information as
the amount of facts, data, knowledge. Different such definitions are proposed in the different versions of Information Theory (Fischer; 1927; Shannon, 1948; Kullback-
Leibler, 1951, etc.). In what follows the attention is
centred on the Shannon information approach. Only, the
indispensable minimum of the mathematical formalism is
given. For more detail the numerous references should be
consulted.

In the Shannon statistical information theory (Shannon
and Weaver, 1949; Brillouin, 1956; Khinchin, 1957; McEl-
ece, 1977) the quantity of information is defined by means
of probability. The probabilistic methods are of use in
situations where there is some uncertainty about the
choice of one or more elements from a certain set. The degree of uncertainty of a given outcome i is expressed by
its entropy H_i as a function of the probability p_i of this
result:

$$H_i = - \ \text{lb} \ p_i \tag{1}$$

(handwritten above: $\log_2$)

When the result is completely determined ($p_i = 1$), the entropy equals zero, whilst in the case of full uncertainty ($p_i = 0$) it is an infinite quantity. (Indeed, $p_i \leqslant 0$, and $\Sigma p_i = 1$).

Another equation, however, is the principal one in the Shannon theory. In this equation the *mean entropy H(P) of the probability distribution* $P = (p_1, p_2, ..., p_k)$ of all the possible outcomes (k in number) in a given situation is defined:

$$H(P) = \sum_{i=1}^{k} \ p_i \ \text{lb} \ p_i \tag{2}$$

Here the logarithm at basis two is taken for measuring the entropy in bits (binary digits).

One should bear in mind that the notion of entropy, as used in Information Theory, is more general than the thermodynamic entropy. Viewed as a measure of the disorder in atomic and molecular motions, the thermodynamic entropy is a specific case of the general notion of entropy as a measure of any kind of disorder, or uncertainty, or uniformity.

Mostly, equation (2) is of use in the description of an experiment in which a given random variable takes k different values having probabilities p_1, p_2, ..., p_k. The quantity of information I is defined as the difference in the entropy values before and after the experiment:

$$I = H(P_o) - H(P_1) \tag{3}$$

Thus, in the statistical information theory of Shannon, *information is measured by means of the reduced uncertainty.* When in an experiment the uncertainty is completely eliminated, the quantity of information equals the initial entropy:

$$I_{max} = H(P_o) = - \sum_{i=1}^{k} p_i \; lb \; p_i \qquad (4)$$

If all P outcomes of the experiment are equiprobable, then

$$H(P) = lb \; P; \quad I = lb \; \frac{P_o}{P_1} \qquad (5),$$

where P_o and P_1 denote the number of outcomes before and after the experiment. Again, in the case of completely eliminated uncertainty (P_1 = 1), the information obtained is a maximum one, and it is equal to the initial entropy:

$$I_{max} = lb \; P_o = H(P_o) \qquad (4')$$

Nonprobabilistic approaches to the quantitative definition of information are also possible. Ingarden and Urbanik (1961) suggested such an axiomatic definition of Shannon's concept of information as a function of finite Boolean rings. The so-called "epsilon-entropy" (an essentially combinatorial quantity), and especially *the algorithmic quantity of information*, both introduced by Kolmogorov (1965, 1969), are of considerable interest. In the second case, the quantity of information is defined as a programme of minimal length, allowing an one-to-one transformation of an object (set) into another. The greater the difference between two objects, the larger the length of the programme, and therefore the latter measures the degree of identity (or diversity) of these objects.

The nonprobabilistic methods in Information Theory extend the notion of quantity of information from a quantity of eliminated uncertatinty to that of *eliminated uniformity* or a *quantity of variety*, in agreement with the

concept of Ashby (1956). It is in such a generalized interpretation that the information-theoretic approaches
are applied to the determination of the so-called *structural information* - the quantity of information contained
in the structure of a certain system.

Every structure is constructed of a certain number of
elements (N) that could be, by means of a selected equivalence relation, partitioned into subsets of equivalent
elements. A finite probability scheme can be associated
with this partitioning:

equivalence classes	$1, 2, \ldots, k$
element partition	$N_1, N_2, \ldots, N_k$
probability distribution	$p_1, p_2, \ldots, p_k$

where $p_i = N_i/N$ is the probability for a randomly chosen
element to belong to the subset i having N_i elements.
Though formally attributed to events (equivalent classes
of elements of a structure) which are not of a stochastic
nature, the probability from the finite probability scheme
meets the Kolmogorov conditions for a probability function
(See Khinchin, 1957). It is zero for an empty set, equals
one for an entirely non-empty set, and it is additive in
case of complementary, mutually exclusive events.

The ratio N_i/N can also be interpreted as the probability for a particular element of the structure to be in-
volved in a given chemical reaction. Thus, the probability
distribution as a whole can be viewed in terms of the
freedom with which the structure will interact with other
structures.

The entropy H(P) of the probability distribution of the
elements of the structure, as defined by eq (2), can be
viewed as a measure of the *mean quantity of information*,
I, contained in each element of the structure:

$$\bar{I} = - \sum_{i=1}^{k} p_i \; lb \; p_i, \; bits/element \hspace{2cm} (6).$$

This statement needs to be discussed in detail. As pointed out by Renyi (1965), H(P) from eq (2) can be interpreted either as a measure of entropy or as a measure of information. The first interpretation is justified when one deals with a system before an experiment is carried out on it. Thus, H(P) measures the uncertainty concerning the results of the experiment. When, however, one deals with a system after an experiment has been carried out on it, H(P) measures the amount of information obtained in the experiment. This is the case with various structures, including chemical ones, like atoms and molecules. The "experiment" reducing the uncertainty of the system is the very process of formation of a structure from the isolated elements. Here information is in bonded from, it is contained in the structure which is the reason to use the term "information content" of the structure. Mowshowitz (1968a) considers additional arguments supporting the statement that the quantity $\bar{I}$, as defined for a certain structure from eq (6) is not a measure of entropy as the term is understood in Information Theory. It does not express the average uncertainty per structure, having N elements, of a given ensemble of all possible structures having the same number of elements. Here, $\bar{I}$ *is rather the information content of the structure under consideration in relation to a system of transformations leaving the structure invariant.*

The specific character of the transformation, determining the kind of structural information, will be discussed in detail in the next chapters. Besides $\bar{I}$, what we shall further call a mean per element information content, the *total information content* (Brillouin, 1956) of the structure will be also of use:

$$I = N.\bar{I} = N \text{ lb } N - \sum_{i=1}^{k} N_i \text{ lb } N_i, \quad \text{bits} \tag{7}$$

Two basic properties of the information functions (6) and (7) should be mentioned here: non-negativity, as well as additivity for independent events.

The concept of structural information based on the interpretation of eq (6) and (7), is consistent with the generalized view on the quantity of information as a quantity of variety (Ashby, 1956). There is no variety in a system composed of identical elements. In that case $\kappa = 1$, $N_i = N$, and $I = \bar{I} = 0$ in eq (6) and (7), i.e. zero quantity of variety and zero quantity of information. In the other extremal case of a maximum variety of structural elements, $\kappa = N$, $N_i = 1$, and the information content of the structure is a maximum one:

$$I_{max} = N \text{ lb } N, \qquad \bar{I}_{max} = \text{ lb } N \tag{8}.$$

It follows from the preceding discussion that Information Theory is applicable to any chemical structure, to which it associates a certain information function and a numerical measure. Some restrictions, inherent to the information approach have, however, to be taken into account. Though precise, the quantitative measures of information are relative but not absolute. They are statistical in nature referring to sets and not to individual elements. Information indices can be associated with various atomic and molecular properties but the link between them is often rather complicated . On the other hand, the large number of information indices, which will be introduced for atoms and molecules in Chapters III, IV and V, puts the question about the theoretical justification of their existence. It should, however, be clearly stated that the proper question is not which of these indices are legitimite and which are not, but whether they are useful and to what extent. The theoretic information indices represent a heuristic approach, born to help in solving some practical questions, and the justification of this approach is in its practical importance.

The large potential applicability of the present information approach is due to the possibilities for various quantitative treatments of chemical structures. The degree of complexity of these structures, as well as their organization and specificity, can thus be compared in a unique quantitative scale. This provides for the rigorous studying of some general features of the chemical structures such as their branching and cyclicity; examining and comparing the degree of organization in different classes of chemical compounds, the specificity of biologically important substances and catalysts, etc.; as well as approaching in a convincing way the question of the degree of similarity or diversity of two chemical objects. The information approach seems fairly convenient for diverse classification problems. It is possible in these cases to derive general information equations for the main groupings of classified objects (nuclides, groups and periods in the Periodic Table of chemical elements, homologous series of chemical compounds, etc.).

The great power of information methods in discriminating similar structures (isomers, isotopes, etc.) is of interest for the computer processing and retrieval of chemical information. These methods are fruitful in discriminating between alternative hypotheses or approximations, the latter being of particular interest for quantum chemistry. Alternatively, the possibilities which Information Theory provides for the creation of new hypotheses are often limited since this theory describes the mutual dependence of a number of variables, but not the behaviour of each of them.

The relation that exists between structure and properties is another field of successful application of information-theoretic approach to chemistry. Evidences on the efficiency of this approach are known on qualitatively different structural levels in chemistry - nuclei, electron shells, molecules, polymers and crystals. Two different paths could be pursued when applying information methods for such purposes. Diverse structural rules can be formulated and these rules can be used without any calculations in predicting the ordering of isomeric compounds with respect to a large number of their properties. Quantitative correlations are also possible between the information indices and properties of chemical compounds and elements. Information indices usually offer an advantage over other structural indices for such correlations since they are capable of expressing the features of chemical structures in full detail. In addition, chemical processes may also be handled by the information approach specifying the change in the information indices upon the interaction.

Probably the most attractive feature of the information approach is its ability to provide a unified view on various phenomena, as well as a universal language for their description. This approach largely extends the interplay between different scientific disciplines allowing useful analogies, as well as common laws, to be devised. Contemporary science tends to unity and Information Theory is one of the promising ways to it. The chemical applications of Information Theory have their place within this tendency, and their importance should be expected to run high in the future.

CHAPTER 3
Atomic Information Indices

1, GENERAL ATOMIC INDICES

An atom represents a system whose structural elements
(protons, p, neutrons, n, and electrons, z) are partitioned
into two substructures: a nucleus of n+p = A nucleons, and
an electron shell of z electrons. According to eq (7), a
definite *atomic information content* (Bonchev and Peev,
1973), corresponds to this atomic structure:

$$I_{at} = (A+z)\ lb\ (A+z) - z\ lb\ z - A\ lb\ A \qquad (9)$$

In fact, eq (9) defines the atomic information contained
in one of the isotopes of a given chemical element. Most
of the chemical elements, however, represent a mixture of
isotopes of different mass numbers A. Only about 20
elements are presented (with an accuracy of up to 0.001%)
by one isotope. Due to this, the *information content of a
chemical element* is defined (Bonchev and Peev, 1973) as
an arithmetic mean from the atomic information of their
isotopes, $I_{at,\ i}$:

$$\bar{I}_{chem.\ elem.} = \sum_i I_{at,\ i} \cdot x_i \qquad (10)$$

where x_i is the relative abundance of isotope i in the
chemical element.

The atomic information indices, as defined by eq (9) and (10), have been calculated for the chemical elements from H to Pb (Table 1). The isotopes which contribute to the total amount of the chemical element by less than 0.001% have been neglected in these calculations.

The atomic information content reflects the periodicity in the properties of the chemical elements. It can also be used for a more complete definition of the information content of molecular structures, especially those of heteroatomic organic compounds (See Chapter V, Section 2).

A more complete definition of the atomic information index is, however, necessary to extent its field of application. This can be done on the basis of the particle distribution into two atomic substructures: the nucleus and the electron shell. Thus, *individual information indices* could be introduced *for the atomic nucleus and the electron shell:*

$$I_{nucleus} = A \text{ lb } A - \sum_{i=1}^{k} p_i \text{ lb } p_i - \sum_{i=1}^{k'} n_i \text{ lb } n_i \quad (11)$$

$$I_{el.shell} = z \text{ lb } z - \sum_{i=1}^{k''} z_i \text{ lb } z_i \quad (12)$$

The number of protons, neutrons and electrons, as well as the mass number, are denoted in eq (11) and (12) by p, n, z, and A, respectively.

The *total atomic information content* is now defined as a sum of eq (9), (11) and (12):

TABLE 1.   Mean information content of chemical elements
1 to 82 calculated on the basis of their most abundant
isotopes

Atomic number	Element	$\bar{I}$	Atomic number	Element	$\bar{I}$	Atomic number	Element	$\bar{I}$
1	H	2.000	28	Ni	78.698	56	Ba	167.880
2	He	5.512	29	Cu	83.071	57	La	170.479
3	Li	8.770	30	Zn	85.749	58	Ce	172.854
4	Be	11.580	31	Ga	89.800	59	Pr	174.970
5	B	14.226	32	Ge	92.979	60	Nd	178.416
6	C	16.544	33	As	95.840	62	Sm	184.957
7	N	19.287	34	Se	99.681	63	Eu	187.806
8	O	22.052	35	Br	101.900	64	Gd	192.041
9	F	25.370	36	Kr	105.777	65	Tb	194.580
10	Ne	27.649	37	Rb	108.251	66	Dy	198.167
11	Na	30.880	38	Sr	111.085	67	Ho	201.150
12	Mg	31.247	39	Y	113.470	68	Er	204.087
13	Al	36.390	40	Zr	116.391	69	Tu	206.730
14	Si	38.623	41	Nb	118.980	70	Yb	210.519
15	P	41.910	42	Mo	122.259	71	Lu	213.233
16	S	44.153	44	Ru	128.393	72	Hf	216.805
17	Cl	47.698	45	Rd	131.120	73	Ta	219.730
18	Ar	51.813	46	Pd	134.989	74	W	222.817
19	K	52.967	47	Ag	137.161	75	Re	225.939
20	Ca	55.176	48	Cd	141.250	76	Os	229.775
21	Sc	59.560	49	In	144.187	77	Ir	232.439
22	Ti	62.824	50	Sn	147.946	78	Pt	235.503
23	V	66.159	51	Sb	151.218	79	Au	238.350
24	Cr	68.420	52	Te	155.980	80	Hg	241.932
25	Mn	71.700	53	I	157.390	81	Tl	245.568
26	Fe	73.860	54	Xe	161.342	82	Pb	248.698
27	Co	77.200	55	Cs	163.920			

$$I_{at.}^{total} = I_{at.} + I_{nucl.} + I_{el.shell} = (A+z) \text{ lb } (A+z) -$$

$$- \sum_{i=1}^{k} p_i \text{ lb } p_i - \sum_{i=1}^{k'} n_i \text{ lb } n_i - \sum_{i=1}^{k''} z_i \text{ lb } z_i \qquad (13)$$

The proton, neutron and electron distribution into different groups, k, k' and k" in number respectively, is carried out by means of diverse criteria. For this reason eq (11) and (12) will be further used for the definition of diverse information indices of the atomic nucleus and the electronic shell.

2. NUCLEAR INFORMATION INDICES

Following the nuclear shell model of Goeppert-Mayer and Jensen, the protons and neutrons in an atomic nucleus can be viewed as being distributed in shells and subshells, as well as according to the values of different quantum numbers: principal, angular momentum, spin, inner, etc. An individual information index is associated with each of these distributions: $I_{nucl.}^{shell}$, $I_{nucl.}^{subshell}$, $I_{nucl.}^{n}$, $I_{nucl.}^{l}$, $I_{nucl.}^{s}$, $I_{nucl.}^{j}$, etc.

The features of the information indices thus defined have been studied in detail (Peev et al., 1972, 1974; Rousseva et al., 1976). They clearly reflect the regularities of the structure of the nucleus, and seem to be a good basis in the search for a broader correlation between nuclear information indices and physical properties.

The importance of the information approach for the study of atomic nuclei can also be demonstrated by an additional nuclear index (Peev et al., 1972; Bonchev et al., 1976 a): the *information on proton-neutron composition of the nucleus*, I_{nucl}^{np}:

$$I_{nucl}^{np} = A \log_2 A - p \log_2 p - n \log_2 n \qquad (14).$$

The mean-per-nucleon value of this nuclear information index can be expressed by an approximate but suffic-
iently accurate equation:

$$\bar{I}_{nucl.}^{np} \approx 1 - \frac{1}{2\ln2} \cdot \frac{\beta^2}{A^2} \qquad (15),$$

where $\beta = n-z$ is the so-called isodifferent number. This
equation is quite similar in form to the expression derived previously for the mean spin information of one electron (Dimov and Bonchev, 1976):

$$\bar{I}_{el.}^{spin} \approx 1 - \frac{1}{2\ln2} \cdot \frac{a^2}{z^2} \qquad (16),$$

where a is the number of unpaired electrons in the electron shell.

Proceeding from the similarity of eq (15) and (16), an
information analogy has been formulated to exist between
nuclides* and chemical elements (Table 2) which might be
of use in the nuclide systematics.

A modification of the information on neutron-proton
composition of a nucleus is defined as the difference of
$I_{nucl.}^{np}$ - values for two nuclei having mass numbers A and
A-1, respectively. It is called *differential information*,
$\Delta I_{nucl.}^{np}$:

---

* According to the definition of IUPAP (1956), the nuclide
is a certain type of atom, or the atom of an isotope
of a certain element.

$$\Delta I^{np}_{nucl} = I^{np}_{nucl} (A) - I^{np}_{nucl} (A-1) \qquad (17)$$

TABLE 2.   Information analogy between nuclides and chemical elements

N NUCLIDES		CHEMICAL ELEMENTS	
Parameters	Groups	Parameters	Groups
1. β = const, $A \neq$ const	Isodifferents	a = const $z \neq$ const	Vertical groups
2. n = const, $\beta \to \beta+1$ $A \to A+1$	Isotones	$z \to z+1$ $a \to a+1$	$(s^1, p^1$ to p^3, d^1 to d^5, f^1 to f^7) Periods
3. z = const, $\beta \to \beta - 1$ z = const	Isotopes	$z \to z+1$ $a \to a+1$	$(s^2, p^4$ to p^6, d^6 to d^{10}, f^8 to f^{14})
4. A = const, β = const $\Delta\beta$ = 2	Isobars	z = const $a \neq$ const Δa = 2	Electronic states of a chemical element with a different spin

This quantity is shown in FIG. 1 as a function of the mass number for the 209 known 2β - stable isotopes. It alters harmoniously approximating 1 bit, forming two branches with positive and negative, respectively, deviations from the mean value of 1 bit. The areas of different type of nucleus filling are set apart in FIG. 1: deutron type (n, p) for $A \leqslant 20$, and helium type (2n, 2p and 4n, 2p) at higher values of the mass number.

The total information of proton-neutron composition of a nucleus is close in magnitude to the mass number A, remaining smaller than, or equal to, it. Their difference:

$$\Delta I_{nucl}^{np'} = A - I_{nucl}^{np} \approx \frac{1}{2\ln 2} \cdot \frac{\beta^2}{A} = \frac{1}{2\ln 2} \frac{(A - 2p)^2}{A} \qquad (18)$$

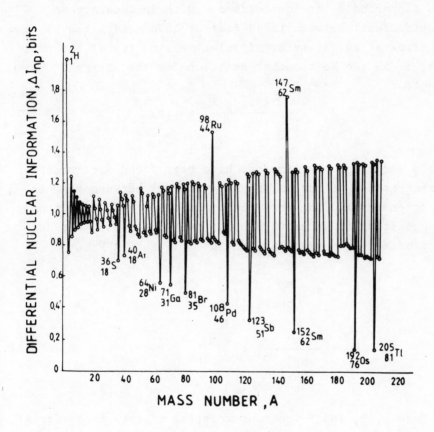

FIG. 1. Dependence of the differential information on proton-neutron composition of the atomic nuclei on their mass number

can be called *"defect" of the information* on proton-neutron composition (Bonchev et al., 1976a), since it expresses the loss of information upon atomic nucleus formation from free nucleons, and recalls the defect of the mass upon the same process. The two defects might be expected to be connected, since they both increase at higher atomic-nucleus deviations from the symmetrical state having an equal number of protons and neutrons. This idea is supported by the coincidence (with an accuracy to a coefficient) between the defect of information (eq 18) and the so-called parameter of relative symmetry of the nucleus, δ, in the Weizsaecker equation for the energy of atomic nucleus

$$\delta = - \varepsilon \frac{(A - 2p)^2}{A}, \text{ energy} \tag{19}.$$

In addition, correlations have been found for isodifferent groups of nuclei ($\beta = n-p = $ const) between the binding energy and the total information on proton-neutron composition of atomic nuclei or, alternatively, the defect of this information:

$$E_b = a. \; I^{np}_{nucl(id)} + b \tag{20}$$

$$E_b = \frac{k}{\Delta I^{np'}_{nucl}} + b \tag{21}$$

Here a, b, and k are constants for a given isodifferent group. The two correlations are illustrated in Table 3 for a number of isodifferent series of nuclides. The correlation coefficients are high particularly for $\beta = n-z > 15$ where the mean relative error is less than 1%.

The information indices on proton-neutron composition of atomic nucleus were also applied to nuclide systematics, introducing a fifth kind of nuclides besides the known

TABLE 3. Constants in the correlations of the nuclear
binding energy with the total information on proton-
neutron composition of atomic nuclei, as well as with
the defect of this information (eq 18,19)*

β	a	b	k	β	a	b	k
5	8.22	27.62	148.9	30	6.42	302.37	4266.9
10	7.95	65.38	579.0	40	5.85	439.21	6977.2
15	7.73	96.16	1275.1	50	5.42	550.61	10096.0
20	6.82	212.07	1999.8	55	5.42	562.32	12252.3

*a - in $meV.bit^{-1}$, b - in meV, k - in meV.bit

four kinds (Chart of the Nuclides, 1963):isotopes (p =
const); isotones (n = const); isobars (A = const); and
isodifferents (β = const).An information-theoretic variant of nuclide systematics was suggested by Rousseva and
Bonchev (1978) on the basis of the mean-per-nucleon defect
of information on proton-neutron nuclear composition,
$i = \Delta I^{np'}_{nucl}/A$. It represents (FIG.2) a diagram of the mean
defect vs. mass number (i/A-diagram) in which each of the
nuclei is placed on the crosspoint of five lines. The new
type of nuclide group includes the nuclei of the chemical
elements having the same mean defect of information. As a
matter of fact these are new lines of nuclear genesis within which each nucleus can be obtained by a reaction of
nuclear fusion between the initial nuclei.

The complete i vs. A-diagram, including all known (Krav-
tsov, 1974) 1723 isotopes of chemical elements, displays
specific features reflecting the fundamental connection
which is supposed to exist between nuclear and electron
structure of the atom.

Thus, in general, the curve of the maximum defect of information (i_{max}) has maxima at the proton magic numbers

(e.g. z = 50, 82). Alternatively, most of the minima on
this curve correspond to the chemical elements whose electron shells are full or half-full: p^6 (z = 10, 18), d^{10}
(z 30, 46, 80), f^{14} (z = 71), p^3(z = 15, 33), d^5 (z = 25,
75).These results recall a similar finding of Lepsius
obtained in a Z vs. β-diagram. Finally, a substantial
increase in the curve of the minimum defect on nuclear
information (i_{min}) occurs for the elements where one
period ends and the next one begins in the Periodic Table.

The analogy in the nuclear and electron properties of
chemical elements, substantiated in detail by means of the
information indices allowed also a good correlation to be
found between the mean defect of information on the proton-
neutron composition i and the electronic energy, E_e, for
isodifferent groups of nuclides (Rousseva and Bonchev,
1979). Equations of the type:

$$lg \ |E_e| = C.i^d \tag{22}$$

$$lg \ |E_e| = \frac{i - i_o}{C + D.i} + lg \ |E_e^o| \tag{23}$$

provided a mean relative error less than 0.1% for elements
from Periods V, VI, and VII, while for periods III and
IV it is within the range 0.1 to 1.0%. Here c, d, C and D
are constants for a given isodifferent number β; i_o and E_e^o
refer to the first nuclide in the isodifferent series.
Similar correlations between electronic energy and nuclear
binding energy allowed the prediction of the latter for 45
nuclides with Z = 101 to 108 (Rousseva and Bonchev, 1980).

It can be concluded that the nuclear information indices
proved to be of use in the characterization of nuclear
structure, to some correlations with nuclear properties,

as well as in establishing the information analogy in the periodic structure of atomic nuclei and electron shells.

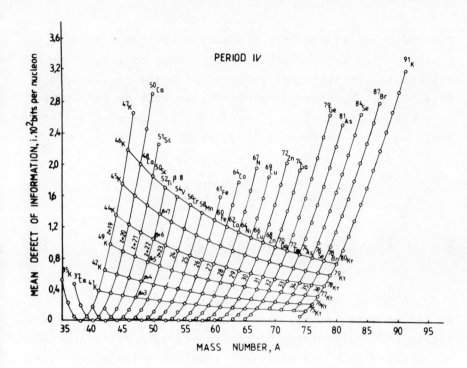

FIG. 2. Dependence of the information defect per nucleon on the nuclide mass number for nuclides of Period 4 (A fragment of the information-theoretic variant of the nuclide systematics)

3. INFORMATION INDICES ON ELECTRON SHELLS OF ATOMS

A. MEAN AND TOTAL INFORMATION INDICES

The electrons in the atomic electron shell could be treated, like the nucleons in the atomic nuclei, as being distributed in different subsets: z $(z_1, z_2, \ldots, z_k)$. On this basis, the mean-per-electron, as well as the total *information on electron distribution in the atomic electron shell* has been introduced (Bonchev et al., 1977):

$$\bar{I}^x_{el.shell} = -\sum_{i=1}^{k} \frac{z_i}{z} \text{ lb } \frac{z_i}{z} \tag{24}$$

$$I^x_{el.shell} = z \text{ lb } z - \sum_{i=1}^{k} z_i \text{ lb } z_i \tag{25}$$

The kind of information index, x, is determined by the criterion selected for electron partitioning into subsets. Within the one-electron approximation such criteria could be selected from the different atomic quantum numbers, as well as from some of their combinations which determine different groups of electrons or atomic energy states. The following types of electron subsets in the atom have been taken into consideration: 1. electron shells; 2. subshells; 3. atomic orbitals; 4. spin-orbitals; 5. nlj-subshells, as well as groups of electrons having the same value of a certain quantum number; 6. angular momentum (1); 7. magnetic (m); 8. magnetic spin (m_s); 9. inner (j); 10. total magnetic (m_j); 11. the sum of the principal and angular momentum quantum numbers (n+1); 12. the sum of the principal and inner quantum numbers (n+j).

Consider, as an illustration of the information approach, the element chlorine having the electronic configuration $1s^2\ 2s^2\ 2p^2\ 3s^2\ 3p^5$. The 17 electrons are distributed in the following subsets: shells (2, 8, 7), subshells (2, 2, 6, 2, 5), nlj-subshells (2, 2, 2, 4, 2, 2, 3), atomic orbitals (8x2, 1), spin-orbitals (17x1); by the values of the quantum numbers: (6, 11) for l = 0,1; (10, 4, 3) for m = 0, $\pm$1; (9, 8) for m_s = $\pm$1/2; (10, 7) for j = 1/2, 3/2; (2, 7, 7, 1) for m_j = - 3/2, -1/2, +1/2, +3/2; (2, 2, 8, 5) for n+l = 1, 2, 3, 4. An information index, determined by each of eq (24, 25), corresponds to each of these 12 distributions. Four of the total atomic information indices of elements 1 - 107 are presented.

Comparing the first five kinds of atomic information indices, one could obtain the following inequalities (for simplicity the electron-shell subscript will be omitted):

$$I_n < I_{nl} < I_{nlj} < I_{nlm} < I_{nlmm_s} \qquad (26).$$

Inequalities (26) show that the more complete definition of electron energy by means of subsequent introduction of additional quantum numbers decreases the uncertainty in energy, and therefore increases the information con - tent of the atom. When electrons are treated as being distributed over spin-orbitals any uncertainty is eliminated and the quantity of information is a maximum:

$$I_{nlmm_s} = N\ lb\ N = I_{max} \qquad (27).$$

The information on electron distribution over spin (or otherwise spin-information), I_{m_s}, manifests interesting

TABLE 4. Mean information of electron distribution over electron shells ($\bar{I}_n$) and atomic orbitals ($\bar{I}_{nlm}$), as well as over the values of the magnetic spin quantum number ($\bar{I}_{m_s}$) and the sum of the principal and angular momentum quantum numbers ($\bar{I}_{n+1}$) for chemical elements Z = 1 to 107

z	$\bar{I}_n$	$\bar{I}_{nlm}$	$\bar{I}_{m_s}$	$\bar{I}_{n+1}$
1	0	0	0	0
2	0	0	1.0000	0
3	0.9182	0.9182	0.9182	0.9182
4	1.0000	1.0000	1.0000	1.0000
5	0.8709	1.5219	0.9709	1.5219
6	0.9183	1.9183	0.9183	1.5850
7	0.8632	2.2360	0.8631	1.5567
8	0.8112	2.2500	0.9544	1.5000
9	0.7642	2.2810	0.9911	1.4355
10	0.7219	2.3219	1.0000	1.3710
11	1.0962	2.5503	0.9940	1.3093
12	1.2516	2.5849	1.0000	1.2516
13	1.3347	2.7773	0.9957	1.5466
14	1.3788	2.9502	0.9852	1.6645
15	1.3996	3.1069	0.9710	1.7282
16	1.4056	3.1250	0.9887	1.7500
17	1.4021	3.1462	0.9975	1.7574
18	1.3921	3.1699	1.0000	1.7527
19	1.6163	3.3006	0.9980	1.7400
20	1.7219	3.3219	1.0000	1.7219
21	1.7004	3.4399	0.9984	1.9161
22	1.6758	3.5504	0.9940	2.0049
23	1.6517	3.6540	0.9878	2.0559
24	1.4972	3.8350	0.9544	2.1158

TABLE 4.   (Continued)

z	$\bar{I}_n$	$\bar{I}_{nlm}$	$\bar{I}_{ms}$	$\bar{I}_{n+1}$
25	1.5996	3.8439	0.9710	2.0995
26	1.5734	3.8543	0.9829	2.1039
27	1.5473	3.8660	0.9911	2.1011
28	1.5216	3.8788	0.9963	2.0931
29	1.3732	3.8925	0.9991	2.0693
30	1.4716	3.9068	1.0000	2.0662
31	1.5408	3.9864	0.9992	2.0492
32	1.5919	4.0625	0.9972	2.0306
33	1.6302	4.1353	0.9940	1.0109
34	1.6590	4.1463	0.9975	1.9903
35	1.6804	4.1578	0.9994	1.9691
36	1.6961	4.1699	1.0000	1.9477
37	1.8294	4.2365	0.9995	1.9260
38	1.9043	4.2480	1.0000	1.9043
39	1.9113	4.3110	0.9995	2.0274
40	1.9150	4.3719	0.9982	2.0955
41	1.8434	4.4794	0.9892	2.1993
42	1.8408	4.5328	0.9853	2.2234
43	1.8365	4.5426	0.9902	2.2403
44	1.8307	4.5503	0.9940	2.2517
45	1.8239	4.5585	0.9968	2.2588
46	1.6949	4.5235	1.0000	2.2797
47	1.8075	4.5758	0.9997	2.2630
48	1.8742	4.5849	1.0000	2.2458
49	1.9234	4.6351	0.9997	2.2451
50	1.9615	4.6839	0.9989	2.2423
51	1.9915	4.7312	0.9975	2.2379
52	2.0153	4.7388	0.9989	2.2320
53	2.0342	4.7468	0.9998	2.2249
54	2.0491	4.7548	1.0000	2.2165

TABLE 4.   (Continued)

z	$\bar{I}_n$	$\bar{I}_{nlm}$	$\bar{I}_{m_s}$	$\bar{I}_{n+1}$
55	2.1429	4.7995	0.9998	2.2078
56	2.1981	4.8073	1.0000	2.1981
57	2.2075	4.8504	0.9998	2.2870
58	2.1770	4.8924	0.9991	2.3387
59	2.1657	4.9334	0.9981	2.3763
60	2.1542	4.9735	0.9968	2.4050
61	2.1421	5.0126	0.9951	2.4270
62	2.1300	5.0510	0.9931	2.4441
63	2.1177	5.0884	0.9918	2.4571
64	2.1299	5.1250	0.9886	2.4669
65	2.0924	5.0991	0.9957	2.4739
66	2.0798	5.1048	0.9974	2.4786
67	2.0669	5.1107	0.9985	2.4814
68	2.0542	5.1169	0.9994	2.4825
69	2.0413	5.1231	0.9999	2.4822
70	2.0285	5.1292	1.0000	2.4805
71	2.0428	5.1637	0.9998	2.4917
72	2.0550	5.1977	0.9994	2.4738
73	2.0649	5.2309	0.9988	2.4693
74	2.0733	5.2635	0.9979	2.4638
75	2.0799	5.2954	0.9968	2.4577
76	2.0853	5.3006	0.9980	2.4512
77	2.0894	5.3057	0.9987	2.4439
78	2.0477	5.3110	0.9996	2.4318
79	2.0491	5.3164	0.9999	2.4229
80	2.0960	5.3219	1.0000	2.4200
81	2.1316	5.3522	0.9999	2.4114
82	2.1610	5.3819	0.9995	2.4023
83	2.1855	5.4111	0.9990	2.3931
84	2.2063	5.4162	0.9996	2.3838

TABLE 4.   (Continued)

z	$\bar{I}_n$	$\bar{I}_{nlm}$	$\bar{I}_{m_s}$	$\bar{I}_{n+1}$
85	2.2238	5.4211	0.9998	2.3740
86	2.2389	5.4263	1.0000	2.3644
87	2.3036	5.4543	0.9999	2.3544
88	2.3444	5.4594	1.0000	2.3445
89	2.3561	5.4870	0.9999	2.4070
90	2.3659	5.5140	0.9996	2.4461
91	2.3537	5.5408	0.9992	2.4762
92	2.3516	5.5670	0.9987	2.5005
93	2.3338	5.5931	0.9979	2.5205
94	2.3301	5.6183	0.9970	2.5372
95	2.3260	5.6434	0.9960	2.5512
96	2.3383	5.6682	0.9950	2.5630
97	2.3339	5.6719	0.9962	2.5726
98	2.3120	5.6555	0.9988	2.5807
99	2.3067	5.6597	0.9994	2.5872
100	2.3012	5.6639	0.9997	2.5925
101	2.2955	5.6681	0.9999	2.5966
102	2.2895	5.6724	1.0000	2.6006
103	2.3022	5.6962	0.9999	2.6018
104	2.3131	5.7197	0.9997	2.6032
105	2.3223	5.6975	0.9994	2.6476
106	2.3301	5.6902	0.9990	2.6400
107	2.3380	5.6797	0.9984	2.6494

features. For all the elements with a closed electron shell (having s^2, p^6, d^{10}, or f^{14} electron configuration) it is exactly equal to the atomic number of the element, i.e. each electron in these elements contributes 1 bit of spin-information:

$$I_{m_s}^{\uparrow\downarrow} = z \text{ bits/atom;} \quad \bar{I}_{m_s}^{\uparrow\downarrow} = 1 \text{ bit/electron} \tag{28}.$$

B. DIFFERENTIAL INFORMATION INDICES

They have been introduced by Bonchev and Kamenska (1978a) by means of eq (29):

$$\Delta I_x = I_{x,z} - I_x, \; z-1 \tag{29},$$

specifying the increase in information content of an atom of a chemical element related to an atom of another element, whose atomic number is smaller by one. The differential indices are more sensitive atomic indices than those discussed above. This can be shown by the following considerations.

The filling of a given electron subset (shell, subshell, etc.) begins and ends most frequently at a constant population of the preceding subsets, k in number. Then eq (29) is transformed to:

$$\Delta I_x = Z \; 1b \; Z - (Z-1) \; 1b \; (Z-1) - (Z_{k+1} \; 1b \; Z_{k+1} -$$

$$Z'_{k+1} \; 1b \; Z'_{k+1} \tag{30}.$$

Evidently, the differential information indices have a
maximum value in every element in which the formation of
a new subset (k+1) begins ($Z_{k+1} = 1$, $Z'_{k+1} = 0$). They decrease regularly with increasing number of electrons in
this subset and have a minimum value when the subset is
maximally filled with electrons. In this way the differential information indices seem to be a convenient means
for the description of the periodicity in the electronic
structure of atoms.

This statement is exemplified in FIG.3 where the differential information on electron distribution over shells
is shown for the chemical elements up to z = 103. This
index has a maximum at the initial element of every period
since for this element the filling of a new electron
shell begins. A curve, common for a given period and ending in a minimum at the corresponding noble gas, is obtained for the second and third period only. In the other
cases the information curve of every period is divided
into different s, p, d, and f-parts, due to the delay in
filling the f- and d-subshells. At this, an inequality
(31) is valid:

$$\Delta I_n^s > \Delta I_n^p > I_n^d > \Delta I_n^f \tag{31}$$

Inequalities (31) show the increase of this information
index in the sequence of f-, d-, p-, and s- subshells.
This result brings to mind the well known difference in
the properties of chemical elements which is the least one
for f- elements and the largest one for s- and p- elements.
This is not a chance coincidence. The properties of chemical elements essentially depend on the effective nuclear
charge. According to the theory of Clementi and Raimondi
the effective charge of an atom depends on the number of

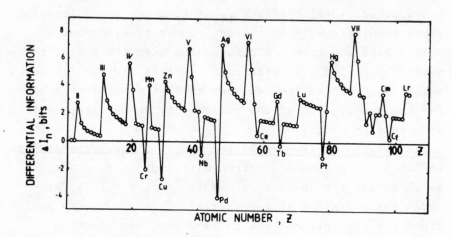

FIG.3. Differential information on electron distribution
over shells in the atoms of the chemical elements

electrons Z, principal and angular momentum quantum numbers
n and l, and atomic quantum numbers on angular - and spin-
momenta, L and S. The first three of these quantities take
part in the determination of the information on electron
distribution over shells and subshells. Hence, information
index ΔI_n will be capable of reflecting the properties of
the chemical elements.

As the type of valence electrons determines the grouping of chemical elements into main, secondary and tertiary (lanthanides, actinides) groups, one can also con-
clude that the information index ΔI_n is capable of expressing correctly the horizontal and vertical structure
of the Periodic Table. In fact, the elements of every
main or secondary group can be connected in FIG.3 in a

common curve, to which a group information equation corresponds. Such equations will be given below in subsection C.

Differing from ΔI_n the differential information index
on the electron distribution over atomic orbitals, ΔI_{nlm},
is a rather monotonous function of the atomic number
(FIG. 4). The total curve consists of two parallel branches. The upper branch includes the elements in which, in
accordance with Hund's first rule, every AO in a given
subshell is populated by unpaired electrons only. These
are one s-, three p-, five d- and seven f- elements. The
lower branch of the differential curve includes the same
number of elements, but from the second half of the subshells, in which every new electron forms an electron
pair. The two branches can be described by eq 30, taking
Z_{k+1} = 1 and 2, respectively. From this one can find that
the difference between them is constant and equals 2 bits.
In such a way the data shown in Figure 4 may be treated
as an illustration of Hund's rule, which demands a maximum
value of ΔI_{nlm}. It may be expected that these data will
correlate with the magnetic properties of atoms.

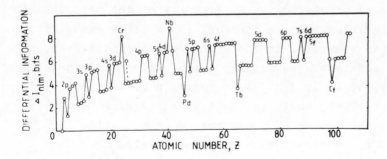

FIG. 4. Differential information on electron distribution
over atomic orbitals in the atoms of chemical elements

C. INFORMATION EQUATIONS FOR THE GROUPS AND PERIODS IN THE PERIODIC TABLE OF CHEMICAL ELEMENTS

The information indices on the electron shells of atoms introduced in subsections A and B have been applied to the study of the quantitative aspects of periodicty.

The first information equation of the groups and periods of chemical elements has been introduced by Dimov and Bonchev (1976). This is the modified equation (16) for the mean electron spin information (the information on the electron distribution over the values of the magnetic spin quantum number, m_s), $\bar{I}_s$:

$$\bar{I}_s = 1 - \frac{a^2}{21n2(Z_o + n)} = 1 - \frac{a^2}{21n2(Z_o + a + b)} \tag{32}$$

Here, Z_o is a constant for a period, equal to the atomic number of the element concluding the preceding period, $n = Z - Z_o$ can be regarded as the sequential number of the element in the period, while a and b are the total number of unpaired and paired electrons respectively in the subshells populated in the elements of the period under consideration; a is a constant for a given group of chemical elements. In the ground state of the atom, the group constant a equals the lowest valence of the element (a = 1, 0, 1, 2, 3, 2, 1, 0 for main groups I to VIII, respectively), while in the highest valence state, it equals the number of the group. Z_o = 0, 2, 10, 18, 36, 54, 86 for period I to VII, respectively.

Three other information equations have been introduced (Bonchev et al., 1976 b) based on the indices for electron distribution over subshells, atomic orbitals, and the values of the angular momentum quantum number. The simp-

lest one is that for the information on electron distribution over atomic orbitals, I_{nlm}:

$$I_{nlm} = (Z_o + a + b) \, lb \, (Z_o + a + b) - Z_o - b \qquad (33)$$

where Z_o, a, and b are the same as defined in relation to
eq (32).

The equation for the information on electron distribution over subshells, I_{nl}, is:

$$I_{nl} = Z \, lb \, Z - A - \sum_{l=0}^{3} K_{nl} \, lb \, K_{nl} \qquad (34).$$

Here the group constant K_{ns} = 1 or 2 for main groups I
and II, K_{np} = 1 to 6 for main groups III to VIII, and
K_{nd} = 1 to 10 for secondary groups I to X. The period
constant A = 0, 2, 19.51, 37.02, 87.75, 138.48, and 242.51
bits for periods I to VII, respectively.

It should be mentioned that some elements having an an-
omalous electron structure like Cr, Cu, Nb, etc. do not
obey eq (32-34) since the number of their unpaired electrons differs from the group constant a.

D. INFORMATION INDICES FOR PERIODS AND SUBPERIODS

Information indices can be directly specified for the
periods and s, p, d, and f-subperiod in the Periodic
Table of chemical elements (Bonchev and Kamenska, 1978 b)
as the total increase of atomic information content in
these groupings of elements:

$$I_x^{period} = \sum_{period} \Delta I_{x,i};$$ (35)

$$I_x^{subperiod} = \sum_{subperiod} \Delta I_{x,i}$$ (36)

where $\Delta I_{x,i}$ is the differential index of the type x for the element i of the period (subperiod) under consideration, as defined by eq (29).

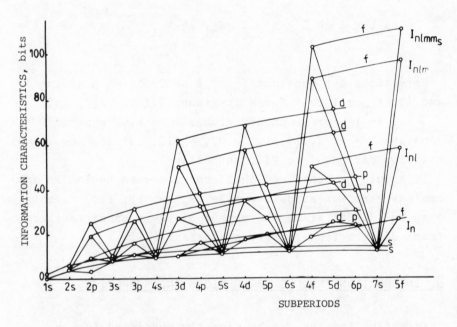

FIG. 5. Four information indices (information on electron distribution over spin-orbitals, I_{nlmm_s}; atomic orbitals, I_{nlm}; subshells, I_{nl}; and shells, I_n) for the subperiods in the Periodic Table of chemical elements

The information indices of all s, p, d, and f-subperiods of the Periodic Table are shown in FIG.5. All sub-

periods of a given 1-type are represented by common approximately parallel lines, the greater 1 is, the higher
these lines lie. A clear regularity exists in the arrang-
ement of periods in ascending characteristic "stages"
each one including a pair of periods. The electron subshells are situated on the right side of the maxima in
FIG. 5 in the sequence $(n-2)f \rightarrow (n-1)d \rightarrow np \rightarrow (n+1)s$, i.e.
following the Klechkovski rules (See e.g. Klechkovski,
1968) for electron subshells filling within the so-called
$(n+1)$-electron groups.

The importance of the $(n+1)$-electron groups is additionally demonstrated in FIG.6 where the differential information index on the electron distribution over these
groups (each group having the same sum of the principal
and angular momentum quantum numbers) is shown. The $(n+1)$-
electron groups are clearly distinguished in this figure
where for each group the information index begins with a
maximum, decreases regularly, and ends with a minimum.
The regular trend of the information function within each
group is violated in a few cases of elements with anoma-
lous electron structures. These deviations from regularity are, however, fewer in number and magnitude, as compared with those from FIG.3, where the distribution of
electrons on shells is considered. The better regularity
displayed by ΔI_{n+1} - index as compared with ΔI_n - index
is viewed as an argument in support of the Klechkovski
ideas, according to which the $(n+1)$ - electron groups are
a more natural system for electron distribution than that
over the $(n, 1)$ - or n-groups, i.e. over subshells and
shells.

The spin-information indices of 4f, 5f, and 5d-subper-
iods, as well as the differential information on electron

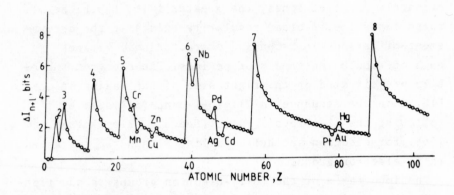

FIG. 6. Differential information on electron distribution
over the sum of the principal and angular momentum
quantum numbers in the atoms of chemical elements

distribution over shells, have been used in the analysis
of the controversial problem of the exact position of the
lanthanides and actinides in the Periodic Table. They
provided evidence in support of the concept of Villar
according to which the lanthanides include the elements
from La to Yb (No 57 - 70) instead of Ce to Lu (No 58 - 71)
and by analogy the actinides from Ac to No (No 89 - 102)
instead of Th to Lr (No 90 - 103). Thus, it is shown that
the spin-information of every period and s-, p-, d- or f-
subperiod should be an integer equal to the number of
elements in it (See also eq 28). This condition is
fulfilled for 4f-, 5d-, and 5f- subshells within the Villar
concept only. On the other hand, in FIG. 3 one finds a
maximum instead of a minimum value of ΔI_n in Lu and Lr;
therefore, Lu and Lr should be regarded as initial elements
in a new subshell (5d or 6d), but not as the last elements
in another subshell (4f or 5f); (See also eq 30 and its
analysis). The presence of a d^1-instead of an f^1-electron
configuration in the initial element of these series (La

and Ac) is explained by $(n-2)f \rightarrow (n-1)d$ electron transition in analogy with other elements of these series
(Gd, Cm, etc.). This view is additionally supported by the
presence of a minimum of ΔI_n at Ce and Pa (FIG. 3),
indicating a restoration of the normal electron configuration, violated by an $f \rightarrow d$ electron transition in La, Ac
and Th.

The atomic information indices allow an opinion to be
offered on another controversial problem in the theory
of Periodic Table: at which element the g-electron will
first appear. The known estimates are within the range
$z = 121 - 126$. A prediction was made (Bonchev et al.,
1977), by extrapolation of the information index for the
electron distribution over the values of the angular
momentum quantum number, that this could occur at the
element having atomic number $z = 122$ or, less probably,
$z = 123$.

E. THE CHANGE IN THE INFORMATION CONTENT OF CHEMICAL ELEMENTS OCCURRING UPON ELECTRON TRANSITIONS

A general analysis has been made (Bonchev and Kamenska,
1979a) of the change in the atomic information content
upon the electron transitions from m-th to n-th electronic subset. Let the electron distributions before and after the electron transition be denoted by R_1 and R_2, respectively:

$$R_1 = z \{z_1, z_2, z_3, \ldots, z_m, z_n\}$$

$$R_2 = z \{z_1, z_2, z_3, \ldots, z_m-1, z_n+1\}$$

$$(37)$$

Let also the respective values of the information indices be I_1 and I_2. Hence, one obtains:

$$\Delta I = I_2 - I_1 = z_m lb\ z_m + z_n lb\ z_n - (z_m-1)\ lb(z_m-1) -$$

$$- (z_n+1)lb\ (z+1) \tag{38}$$

where

$$\Delta I = 0 \qquad \text{for} \qquad z_m = z_n+1 \tag{39a}$$

$$\Delta I > 0 \qquad \text{for} \qquad z_m > z_n+1 \tag{39b}$$

$$\Delta I < 0 \qquad \text{for} \qquad z_m < z_n+1 \tag{39c}$$

Proceeding from the fact that $n = 1$ for a singly excited state the following conclusion has been made:

The light emission of an atom is associated with a decrease, while the light absorption with an increase in the atomic information content:

$$\text{Emission: } \Delta I \leqslant 0; \text{ Absorption: } \Delta I \geqslant 0 \tag{40},$$

where the equalities hold only for electron transitions from singly occupied to unoccupied energy levels. Proceeding from the concept of information as a negative component of entropy, as well as taking into account eq (40), it has also been concluded that emission is accompanied by an increase in entropy, and therefore it is an irreversible process. This conclusion is consistent with the results of Andrade e Silva and Lochak, obtained by means of thermodynamics of an isolated particle by de Broglie. Therefore, though obtained within the one-electron approximation, eq (40) seems qualitatively correct.

The change in the atomic (or molecular) information content upon laser excitation is not straightforwardly expressed. Still, an equality exists for the pairs of electronic states with the normal and the respective inverse population. Hence, the pairs of atomic (as well as molecular) states with equal positive and negative absolute temperature are characterized by equal information content:

$$I_X(+T, K) = I_X(-T, K) \tag{41}$$

F. CORRELATIONS BETWEEN ATOMIC INFORMATION INDICES AND PROPERTIES OF CHEMICAL ELEMENTS. PREDICTING THE PROPERTIES OF THE 113-120 TRANSACTINIDE ELEMENTS

The great capability of the Periodic Table to predict properties of chemical elements has been known since the time of Mendeleev. Numerous correlations have been obtained in which a certain property of the chemical elements in a group or period is expressed as a function of the atomic number or the period number. The latter two numbers, however, are equal to the total number of electrons in the atom, and the number of electron shells, respectively. Thus correlations derived and extrapolations made for the properties of the superheavy elements are not based on a *detailed* description of the electronic structure of atoms. The information indices introduced for the description of atoms seem to provide a better basis for the prediction of structure-dependent properties of chemical elements since they reflect the electronic structure of atoms in detail.

This idea has been developed by Bonchev and Kamenska (1981) who have used the atomic information indices for predicting various properties of elements 113-120 which belong to the main groups III - VIII of period VII, and main groups I and II of period VIII. The problem for predicting properties of transactinide elements became of importance after the prediction of islands of nuclear stability around elements 114 and 164 which prompted greatly the efforts to synthesize new superheavy elements as well as to search for some of them in nature.

The procedure developed by Bonchev and Kamenska (1981) for the prediction of the properties of elements 113-120 includes correlating the properties and information indices for each of main groups I to VIII and extrapolating to the superheavy element of interest. Four information indices: I_n, $\bar{I}_n$, I_{nl}, and $\bar{I}_{nl}$ (denoted in Table 5 by I_1, I_2, I_3, and I_4, respectively) and, for the sake of comparison, the atomic number z have been used in the correlations. Each of the above five correlations has been obtained by least-square fitting to eight different versions of trial equations. The latter are expected to be mainly different power or exponential type of functions, due to the logarithmic dependence between the atomic information indices and the number of electrons in the atom and its electron subsets:

$$y = A + Bx \tag{42}$$

$$y = A + Bx + Cx^2 \tag{43}$$

$$y = Ax^B \tag{44}$$

$$y = Ax^B + C \tag{45}$$

$$y = 10^{Bx}A \tag{46}$$

$$y = 10^{Bx}A + C \tag{47}$$

$$y = \frac{x}{A + Bx} \tag{48}$$

$$y = \frac{x - x_1}{A + Bx} + y_1 \tag{49}$$

Thus, the best group correlation for a given property has been selected out of 40 equations as that which displays the lowest mean relative error.

The best correlations for 12 properties or atomic characteristics of the chemical elements of main groups I, V, and VIII are shown in Table 5. Every entry contains the type of equation (eq 42 - 49 are marked as 1-8, respectively), then the variable used (the type of information index I_1 to I_4, or the atomic number z), followed by the mean relative error in percent. Coefficients A, B, and C from eq 42-49 are presented in the next three lines. For more details, as well as for the correlations for the remaining main groups the reader should consult the original paper of Bonchev and Kamenska, 1981.

Analysis of Table 5 reveals that the correlations with the information indices substantially improve on those made with the atomic number of chemical elements (66 against 23 cases). The decrease in the relative error achieved when replacing the correlations with the atomic number by those with the information indices is in some cases very impressive: S^o_{solid} (group IV), 0.9% instead of

TABLE 5. Correlations between the atomic information indices I_i (or atomic number z) and some properties of elements of groups I, V, and VIII

Properties	Group I	Group V	Group VIII
S^o_{gas},	3, z; 0.17	3, I_2; 0.09	3, z; 0.29
cal deg^{-1}(g-atom)$^{-1}$	30.17	31.67	28.52
	0.08218	0.0604	0.08805
S^o_{solid},	4, I_4; 3.15	8, I_2; 0.33	
cal deg^{-1}(g-atom)$^{-1}$	0.1773	13.86	
	3.181	0.06230	
	11.56		
H_{melt},	8, z; 0.76	5, I_2; 0.11	2, I_3; 3.71
kcal(g-atom)$^{-1}$	-0.9169	9.685	0.0500
	-1.900	-0.001847	-0.05426
			0.1495
H_{subl},	5, I_3; 2.27	8, I_3; 0.63	4, I_4; 1.24
kcal(g-atom)$^{-1}$	38.55	0.01247	0.01913
	-0.1577	0.01003	3.892
			1.205
T_M, K	8, I_2; 0.35	5, I_2; 2.64	6, I_3; 2.14
	-0.07876	1444	35.16
	-0.006144	-0.001367	0.3723
			-311.0
T_B, K	8, I_2; 0.71		6, I_4; 2.50
	-0.007112		39.72
	-0.001447		0.2221
			-312.8
I_1, eV	3, I_3; 1.42	3, z; 4.31	8, I_2; 1.91
	5.232	22.73	-4.046
	-0.3611	-0.2520	-0.05752

TABLE 5.   (Continued)

Properties	Group I	Group V	Group VIII
I_2, eV	3, I_4; 2.66 70.39 -0.8829	8, I_2; 3.70 0.7276 -0.8914	3, I_3; 1.32 33.74 -0.6040
V_A, $cm^3 (g\text{-}atom)^{-1}$	4, I_3; 2.37 13.47 2.009 8.181	6, I_1; 6.04 357.3 0.00004479 -344.0	2, I_3; 4.75 19.43 -17.08 13.24
ρ_R, $g\ cm^{-3}$	1, z; 3.6 0.3769 0.02945	7, I_1; 10.1 6.115 0.0723	4, I_2; 1.88 89.91 0.01390 -92.92
R_{cov}, Å	3, z; 3.30 0.9944 0.2147	8, I_3; 0.89 0.7119 0.4495	
ξ	3, I_2; 1.34 0.5704 0.1197	5, I_1; 3.78 1.735 0.0005945	4, I_1; 1.06 0.002824 0.9714 2.212

6.5%; R_{cov} (group VII), 0.14% instead of 3.2%; V_A (group I), 2.4% instead of 10.6%; ΔH_M (group VIII), 3.7% instead of 17.9%, etc. Still greater superiority of the information indices has been found in a preceding study (36 against I cases) where polynomial type functions were solely examined in the correlations (Bonchev and Kamenska, 1979b). The atomic number proved to be of greater importance only for the density and entropy in the gas phase (5 out of 8 group correlations). One could, however, expect the systematic examination of the other eight information indices, mentioned in the Subsection 3. A, to demonstrate also in the remaining cases the advantage of information indices for group correlations in the Periodic Table. Being detailed and flexible characteristics of the electronic structure of atoms, the information indices are capable of describing more adequately than the atomic number or row number the structure-dependent properties of the chemical elements.

The predicted values for 12 macroscopic properties or atomic characteristics of elements 113-120 are shown in Table 6. In general they are consistent with the predictions made by other authors and substantially supplement them.

Naturally, making use of a greater variety of indices and functions for the correlations, some of the expected values presented in Table 6, could be further improved. An additional refinement of the extrapolation procedure may come from modification of the atomic information indices so as to take into account the major role of the outermost electrons in the chemistry and including them with larger weights in the information functions.

The atomic information indices can also be applied to correlations with the properties of some series of chemical compounds. Linear correlations between each of the twelve information indices and the enthalpy of formation of alkali metal fluorides and lithium halides has been found by Bonchev et al. (1975), e. g.:

$$- (\Delta H)_{MeF} = - 4.677 \, I_{nlm} + 150.2 \tag{50}$$

with a standard deviation of s = 0.908, and correlation coefficient r = 0.995.

TABLE 6.   Predicted properties of elements 113-120

Pro-perties[a]	113	114	115	116	117	118	119	120
S^o_{gas}	44.5	43.4	45.8	46.2	45.8	43.4	44.7	43.4
S^o_{solid}	17.1	18.8	15.4	17.0	-	-	25.7	19.0
ΔH_M	1.26	-	1.42	1.82	-	0.83	0.48	2.08
ΔH_S	32.4	41.7	58.5	28.5	-	5.20	15.4	41.0
T_M	727	303	349	708	767	255	296	960
T_B	1395	1700	-	1085	842	259	935	1830
I_1	6.00	7.9	6.9	7.80	8.57	9.48	3.74	5.47
I_2	18.9	16.4	21.3	18.7	19.2	19.5	20.8	9.40
V_A	18.3	20.7	23.8	30.1	44.4	58.3	94.7	45.2
ρ	15.8	13.7	12.7	11.2	7.2	5.0	3.8	5.2
R_{cov}	1.76	1.74	1.57	1.64	1.56	-	2.72	2.08
ξ	2.18	2.17	2.34	2.51	2.72	2.91	1.20	1.37

[a]Same dimensions as in Table 5

G. INFORMATION THEORETIC INTERPRETATION OF THE PAULI EXCLUSION PRINCIPLE AND THE HUND RULE

Some   generalizations have also been made on the basis
of the atomic information indices (Bonchev, 1981). Thus
the first Hund rule, governing the initial population of
the degenerate atomic (as well as molecular) orbitals by
unpaired electrons of the same spin appears to be associated with the requirement for a maximum information on
the electron distribution over the atomic (or molecular)

orbitals. For instance (See also FIG.4):

$$I_{AO(MO)} = max \qquad I_{AO(MO)} < max$$

The Pauli exclusion principle, which does not allow a spin-orbital to be populated by more than one electron, has been interpreted in a similar manner. As already shown by eq (27), maximum information on electron distribution over spin orbitals follows from this restriction. Thus, the Pauli exclusion principle is related to a trend towards acquiring maximum information in atoms and molecules, and more generally, in any fermion system. This interpretation has been additionally supported by an information analysis of electronic wave functions (both the Hartree-Fock and the multideterminantal ones), as well as of the irreducible representations of the symmetry group S_N. The last case is given below in more details.

As known, every system having N indistinguishable particles belongs to the S_N symmetrical group whose order is equal to N. The particles are divided into two groups, bosons and fermions, with respect to their behaviour towards the permutation operations. The wave function of bosons remains unchanged upon all possible permutations of the particles. Contrary to this, the wave function of fermions reverses its sign upon permutations having an odd number of transpositions, and remains unchanged only for an even number of transposition permutations.

As an illustration a part of the character table (Table 7) of the S_N symmetrical group is given. The irreducible representations to which the boson and fermion wave

TABLE 7. Character table of the $\lceil [N]$ and $\lceil [1^N]$ irreducible representations of the S_N symmetrical group

Permutations irreducible representations	$(N!/2)P_0$	$(N!/2)P_1$
$\lceil [N]$	+ 1	+ 1
$\lceil [1^N]$	+ 1	- 1

functions belong are denoted in Table 7 by $\lceil [N]$ and $\lceil [1^N]$, and the even and odd transposition permutations are denoted by P_0 and P_1, respectively. P_0 and P_1 permutations form two subsets of permutations, each of which contains half of the total number of the permutation operations $N!/2$.

As seen, all the characters of the irreducible representation $\lceil [N]$ are the same, while for $\lceil [1^N]$ they are of two different types. Hence, the probability for a system of N bosons to have its wave function unchanged upon a permutation of two bosons is $p_{+1} = 1$, while for a system of N fermions the probability to change or not to change its wave function upon a permutation of two fermions is one and the same: $p_{+1} = p_{-1} = \frac{1}{2}$. Then, applying eq (6) one obtains the amount of information for the symmetry behaviour of system of N fermions and a system of N bosons, upon a permutation of two particles of the same type, respectively,

$$\bar{I}_{fermion} = -\frac{1}{2} \, lb \, \frac{1}{2} - \frac{1}{2} \, lb \, \frac{1}{2} = 1 \text{ bit} = max, \qquad (51)$$

$$\bar{I}_{boson} = 0 \text{ bits} = \min. \tag{52}$$

Clearly, every permutation of two bosons will not give any information, since there is no variety in the symmerry behaviour of the boson wave function. Conversely, every permutation between the fermions will give the maximum possible quantity of information which is exactly 1 bit.

One could suppose that the Pauli exclusion principle and the first Hund rule are specific cases of a certain more general *principle of maximum information content of fermions and zero information content of bosons.* Other known or unknown rules or laws might also be associated with this general principle.

Thus, for instance, the well-known fact that two fermions cannot simultaneously exist in one and the same place and time, though related to the Pauli principle, may independently be interpreted in terms of the space-time distribution within a certain system. One arrives in this way at subsets of cardinality $N_i = 1$, and according to eq (6) and (7), at a maximum of the information on the space-time distribution of elements within any fermion system. Conversely, if the mutual penetration of two objects is possible, then at least one $N_i > 1$, and the information on the space-time distribution of fermions would not be maximal.

Since every electron in an atom or molecule, as well as every fermion system, is in a different spin orbital, following the Ashby definition of information (Ashby, 1956) one may also state that the Pauli exclusion principle is a principle for acquiring maximum variety in atoms and molecules, or more generally in fermion systems.

Finally, it should be mentioned that the main duty of information in cybernetics is to serve in governing var-

ious systems. However, in order to govern a particular
system its elements have to be distinguishable. Thus, new
aspects of the Pauli principle are revealed. Considered
as a principle of acquiring maximum variety in fermion
systems, the Pauli principle makes possible the governing
in them, and first of all in atoms and molecules. This
could be of importance for the expected overlap between
cybernetics, physics, and chemistry.

H. CONCLUDING REMARKS

Concluding this section an opinion can be expressed that,
as a whole, the concept of information content of chemical elements is a convenient mathematical model which reflects the real properties of the chemical elements and
is likely to be homomorphic to the Periodic Table. No other mathematical function could manifest in such a dramatic
way the difference in the properties of chemical elements,
as well as their periodicity. The information approach
could be of use in the analysis of the quantitative aspects of periodicity, in the search for correlations with
the properties of chemical elements and compounds, as a
means in solving controversial problems in the theory of
atomic structure and the Periodic Table, in the estimates
of molecular complexity, etc. The supposed principle of
maximum information content could prompt the search for
new rules of regularities in chemistry.

CHAPTER 4
Information Indices for
Molecules

The application of Information Theory to molecular entities dates back to the early fifties when information theoretic considerations of some biologically important molecules, as well as of their role in the living organisms, have been made for the first time. The methods known for estimation of the molecular information content and their possible applications are presented in this chapter. Different criteria for equivalency of atoms in a molecule are examined: chemical identity, possible ways of bonding through space, molecular topology and symmetry, etc. Special attention is paid to the topological information indices which are the most developed field of application of molecular information indices. Some indices for electron distributions in molecules are also presented.

1. INFORMATION ON ATOMIC COMPOSITION (KIND OF ATOMS)

This is the first information index introduced by Dancoff and Quastler (1953) as an "information for the kind of atoms in a molecule".

Suppose a molecule is denoted by an empirical formula $A_x B_y C_z$. Eq (6) and (7) allow the total and mean-per-atom *information indices on atomic composition* to be calculated in bits:

$$I_{AC} = (x+y+z) \ \mathrm{lb} \ (x+y+z) - x \ \mathrm{lb} \ x - y \ \mathrm{lb} \ y - z \ \mathrm{lb} \ z \quad (53)$$

$$\bar{I}_{AC} = - p_x \ \mathrm{lb} \ p_x - p_y \ \mathrm{lb} \ p_y - p_z \ \mathrm{lb} \ p_z \quad (54)$$

where the probability p_x, p_y, and p_z of a randomly chosen atom to be of kind A, B, and C, respectively, is equal to the relative content of the elements in the molecules: $p_x = x/ \ (x+y+z)$; $p_y = y/ \ (x+y+z)$; $p_z = z/ \ (x+y+z)$. For instance, in the case of C_4H_5Cl, these probabilities are $p_C = 0.4$, $p_H = 0.5$, and $p_{Cl} = 0.1$, respectively. Another example are the living things which are built from 64 different chemical elements. Proceeding from their relative content: hydrogen, 65%; oxygen, 22%; carbon, 7%, etc. one obtains $p_H = 0.65$, $p_O = 0.22$, $p_C = 0.07$, etc., therefrom $I_{AC} \approx 1.5$ bits per atom:

For some classes of organic compounds the information on atomic composition can be expressed as a simple function of the number of carbon atoms, n:

annulenes, C_nH_n:

$$I_{AC} = 2n \ \text{bits}, \ \bar{I}_{AC} = 1 \ \text{bit} \quad (55)$$

cycloalkanes, C_nH_{2n}:

$$I_{AC} = 3n \ \mathrm{lb} \ 3n - n\mathrm{lb}n - 2n \ \mathrm{lb} \ 2n = 3n \ \mathrm{lb} \ 3 - 2n =$$
$$= 2.7549n \ \text{bits} \quad (56)$$

$$\bar{I}_{AC} = 0.9183 \text{ bits} \tag{57}$$

polyenes, $C_n H_{n+2}$:

$$I_{AC} = (2n+2) \text{ lb } (2n+2) - n \text{lb} n - (n+2) \text{ lb } (n+2) \tag{58}$$

alkanes, $C_n H_{2n+2}$:

$$I_{AC} = (3n+2) \text{ lb } (3n+2) - n \text{lb} n - (2n+2) \text{ lb } (2n+2) \tag{59}$$

Dancoff and Quastler (1953), and Zemanek (1959) made use of the information indices for atomic composition in the estimation of the total information content of various organisms, including a living cell ($I \approx 10^{11}$ bits) and man ($I \approx 10^{25}$ to 10^{28} bits). The possible applications to some correlations with molecular properties, as well as to the estimations of molecular complexity will be illustrated in the next sections.

2. INFORMATION ON THE WAYS OF LINKING ATOMS IN A MOLECULE (KIND OF BONDS)

In 1955 Morowitz introduced the first information index which took into account some structural features of molecules. This approach combines the *information* on atomic composition and that *on the possible valence bonds* between the atoms in a system, I_{PB}:

$$I_{MOR} = I_{AC} + I_{PB} \qquad (60)$$

$$I_{PB} = \sum_{i=1}^{k} N_i \; 1b \; L_i \qquad (61)$$

where N_i is the number of atoms of kind i and L_i is the number of bonds which can be formed by an atom of kind i. Upon the determination of L_i one deals with the atoms in the system as being distributed in cubic cells. Thus, every atom may have no more than six neighbours (FIG. 7).

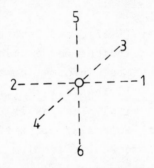

FIG.7. Six possible directions for chemical bond formation in the Morowitz method

This results in six possible bonds for the elements of valence one ($L_1 = 6$). At higher valences the value of L is higher (FIG. 8a, 8b, and 8c).

In the general case the number of all possible directions of the valence bonds is specified as the number of combinations of n elements of class r taken r at a time with repetitions:

$$L_i = C'_{n,r} = \left(\begin{array}{c} n+r-1 \\ r \end{array} \right) = \frac{(n + r - 1)!}{r! \, (n - 1)!} \qquad (62),$$

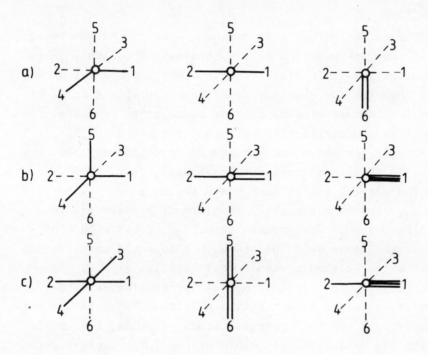

FIG. 8. Some possible chemical bond directions for element of valence: a) two; b) three; c) four

where $n = 6$, and r is the valence of the atom. The following values of L_i have been obtained for elements of a valence 1 to 6:

$$L_{i,1} = 6, \; L_{i,2} = 21, \; L_{i,3} = 56, \; L_{i,4} = 126, \; L_{i,5} = 252,$$

and $L_{i,6} = 462$, e.g. $L_H = 6$, $L_O = 21$, $L_{N,3} = 56$,

$L_{C,4} = 126$, etc.

The method of Morowitz has been widely used in biology for the determination of the information content of organisms in static state, and in diverse processes as well. Of interest for chemistry are the calculations of information content of some classes of organic compounds, as well as the correlation between information indices, heats of formation and entropy of formation of these compounds (Valentinuzzi and Valentinuzzi, 1962, 1963).

The careful analysis of the Morovitz information index, I_{MOR}, reveals a certain inconsistency between its two terms (eq 60). The second term, I_{PB}, deals with the *possible* bonds that the atoms in a molecule may form but not with the kind of bonds that actually exist in it. Conversely, the first term (the information on atomic composition) deals with the kind of atoms that form the molecule, not with all possible kinds of atoms, i.e. not with the whole Periodic table of chemical elements. One should remember, however, that Morovitz developed his method aiming at some quantitative order-disorder considerations in living systems, not in molecular systems.

In the case of molecules, one can proceed from the different nature of the bonds, formed in the molecule under consideration:

$$I_B = B \text{ lb } B - \sum_{i=1}^{m} B_i \text{ lb } B_1 \qquad (63)$$

where B and B_i are the total number of valence bonds and
the number of valence bonds of type i respectively. Such
an information bond index, I_B, has been used by Dosmorov
(1982) in his general equation for the total information
content of a molecule (See Chapter V) though the criterion
used for the bond classification has not been reported.
The simplest version of eq (63) is obtained when the valence bonds are regarded as belonging to one of the following classes: single, double, triple, and aromatic
bonds. As an illustration the molecule of styrene having
B = 16 bonds is considered:

Here $B_1 = 9$, $B_2 = 1$, and $B_3 = 6$ for the single, double,
and aromatic bonds, respectively. Using eq (63) one arrives at the value $I_B = 19.98$ bits.

More detailed bond distributions in molecules will be
dealt with in subsections 3 to 8 on the basis of graph-
theoretical and quantum-mechanical considerations.

Dosmorov (1982) also proposed a modification of eq (63):

$$I_{B'} = B' \text{ lb } B' - \sum_{i=1}^{m'} B_i' \text{ lb } B_i' \qquad (64)$$

so as to comprise also the nonvalent interactions (bonds) in the molecule.

3. INFORMATION ON THE TOPOLOGICAL STRUCTURE OF MOLECULES (TOPOLOGICAL INFORMATION INDICES)

A. REMARKS ON THE GRAPH-THEORETICAL CHARACTERIZATION OF MOLECULES (TOPOLOGICAL INDICES)

i. Introduction

It is the conviction of many chemists nowadays that "the time is ripe for chemistry to become more than a collection of compounds, properties and reactions, namely a coherent unique logical system", as stated by Balaban (1975). Graph theory (Harary, 1969 ; Behzad and Chartrand, 1972; Wilson, 1972) is a basic tool for everybody who shares the above goals. It offers a variety of concepts and methods of importance for chemistry (Balaban, 1976, 1980; Gutman and Trinajstić, 1973; Polansky, 1975; Randić, 1979; Rouvray, 1971, 1973, 1974, 1975 a; Rouvray and Balaban, 1979; Trinajstić, 1982; etc.). Graph theory helps in solving many chemical problems such as the systematization and enumeration of chemical compounds, their coding and nomenclature, correlation of properties, molecular design, automated structural formula search in mass spectrometry, infrared spectroscopy, NMR, etc. Chemical kinetics, phase equilibria, and, last but not least, chemical technology are also subjects of the chemical applications of graph theory.

A *graph G* is defined as a finite non-empty set of V (G) of N *vertices* (points) together with a set E (G) of *edges*

(lines), the latter being unordered pairs of distinct
vertices. Thus, by definition, every graph is finite and
has no loops (an edge initiating from and ending in one
and the same vertex) and multiple edges. When two vertices x and y are joined by an edge e = {x,y}, vertices x
and y are said to be *adjacent* and each of them is *incident*
with the edge e. If the definition of a graph is generalized so as to include multiple edges, a *multigraph MG* is
obtained. A *directed graph* (or digraph) *DG* consists of a
finite set V (DG) of vertices together with a set E (DG)
of *ordered pairs* of distinct vertices of V (DG). The
ordered pairs of such vertices (x,y) are called *directed*
edges or *arcs* (FIG.9)

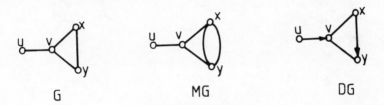

FIG. 9. A graph G, multigraph MG, and a directed graph
   DG

As mentioned by the Nobel-prize winner Prelog (Balaban,
1976) "for more than a century most chemists have used
constitutional formulae without realizing that by representing "connectedness" of atoms such formulae are graphs
or multigraphs". As a matter of fact, the structural
(constitutional) formula of a chemical compound may be
regarded as a *molecular graph* where the vertices represent
atoms while the edges represent valence bonds. Other graphs
are also of great importance for chemistry like *reaction
graphs* (Balaban et al., 1966), corresponding to reaction

mixtures, where vertices symbolize chemical species and edges symbolize conversions between these species; *kinetic graphs* (Yatsimirsky, 1973), where vertices denote intermediate reaction species and edges denote elementary reaction steps, etc. Dealing with molecular graphs *hydrogen depleted* (or skeletal) *graphs* are usually taken into consideration (FIG.10). Unless otherwise stated, these will be the molecular graphs of use in this book.

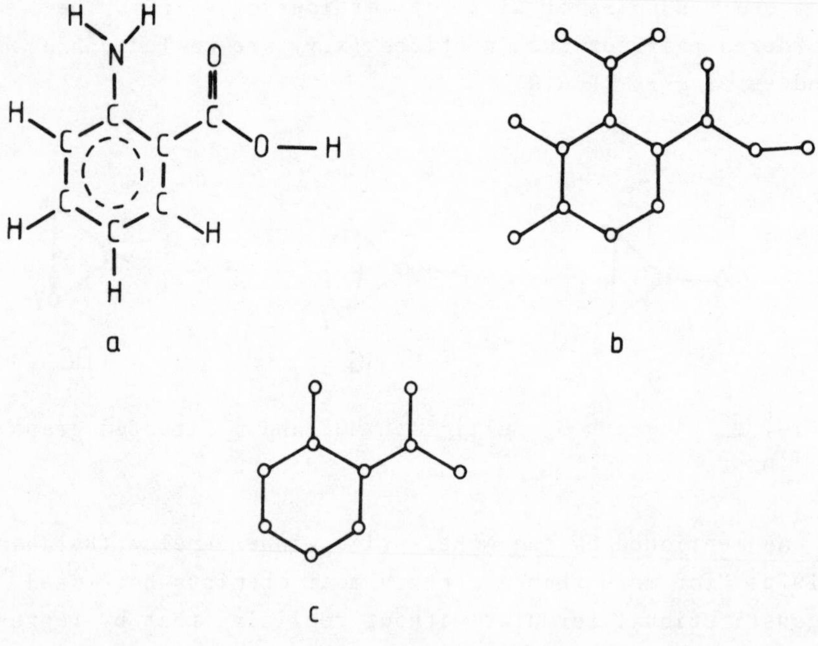

FIG. 10.   o-Aminobenzoic acid: a) structural formula; b) molecular graph; c) hydrogen-depleted molecular graph

The major types of molecular graphs are illustrated in
FIG. 11

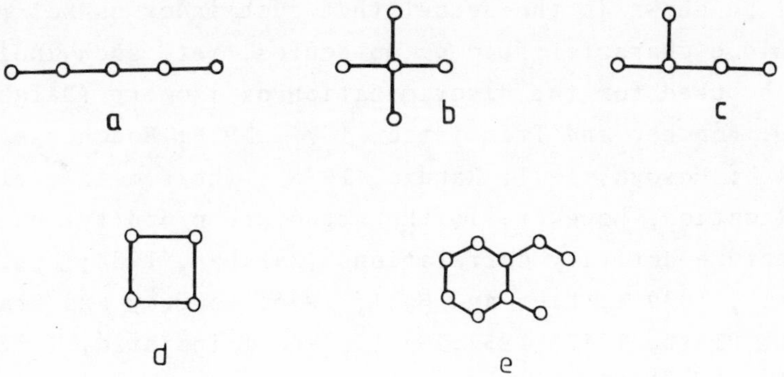

FIG. 11. Molecular graphs: acyclic or tree-graphs (a-c)
and cyclic graphs (d,e); chain (a); star (b); tree
(c); cycle (d); cycle with branches (e)

The graph theoretical characterization of molecular
structure is most often made by its translation into molecular descriptors of two kinds: chemical codes and topological indices. The *chemical code* is a numerical, or
alphabetic or alphanumerical sequence which is uniquely
derived from the molecular structure. It is applied to
computer storage, processing and retrieval of chemical information, providing fast searching for chemical structures or their fragments. The classical codes like the
Dyson's or IUPAC system and the Wiswesser line notation are
not constructed on a graph-theoretical basis, contrary to
the later codes developed by Lederberg (1969), Read
(1978, 1980), etc.

By a topological index, a real number characterizing the
molecule is meant (Balaban et al., 1980, 1982; Bonchev et
al., 1979 a; Rouvray and Balaban, 1979; Trinajstić, 1982).

It is based on a certain topological feature of the corresponding molecular graph and represents a graph-invariant, i.e. does not depend on the vertex numbering. It will be shown in the sequel that such index cannot provide a unique characteristic of molecules. Yet, such indices can be used for the discrimination of isomers (Balaban, 1979; Bonchev and Trinajstić, 1977, 1978; Bonchev et al., 1981 b; Hosoya, 1971; Randić, 1975 . Their main field of application, however, is the structure-property and structure-activity correlations (Balaban, 1980; Bonchev et al., 1979 e; Kier and Hall, 1976; Rouvray and Crawford, 1976; Platt, 1947, 1952; Sablić and Trinajstić, 1982; Smolenski, 1964).

Different graph characteristics or invariants have been used in the definition of molecular topological indices. We shall briefly outline here the major topological indices since they usually originate from the same graph characteristics as the topological information indices and they are developed for the same practical reasons.

ii. The Zagreb Index, M (G)

The number of edges incident with a vertex i in graph G is called its *vertex degree,* v_i. A simple topological index, M, has been defined on this basis by Gutman et al. (1975):

$$M\ (G) = \sum_{i=1}^{N} v_i^2 \qquad (65)$$

where N is the total number of vertices in the graph (atoms in the molecule).

Examples:

$$v_1 = v_5 = v_6 = v_7 = v_8 = 1$$
$$v_4 = 2, v_3 = 3, v_2 = 4$$
$$M (G_1) = 5.1^2 + 2^2 + 3^2 + 4^2 = 34$$

$$v_1 = 1, v_4 = v_5 = 2;$$
$$v_2 = v_3 = v_6 = 3$$
$$M (G_2) = 1^2 + 2.2^2 + 3.3^2 = 36$$

iii. The Randić Connectivity Index, x_R (G)

Randić (1975) applied the vertex degrees to the definition of the so-called *molecular connectivity* index, x_R:

$$x_R(G) = \underset{\text{edges}}{\Sigma} (v_i v_j)^{-1/2} = \underset{\text{edges}}{\Sigma} (x_{v_i} x_{v_j}) \qquad (66)$$

where v_i and v_j are the degrees of a pair of neighbouring vertices *i* and *j* forming the edge {i, j}, and the summation is over all edges in the graph. $x_{v_i v_j}$ may be called *edge* (or partial) *connectivity*. The number of different edge connectivities is limited. In the case of organic compounds with carbon skeletal graphs, where only four types of carbon atoms (primary, secondary, tertiary, and quaternary) exist, there are only nine types of edge connectivities:

$$x_{12} = (1 \times 2)^{-1/2} = 0.7071; \quad x_{13} = (1 \times 3)^{-1/2} = 0.5774;$$

$$x_{14} = (1 \times 4)^{-1/2} = 0.5000; \quad x_{22} = (2 \times 2)^{-1/2} = 0.5000;$$

$$x_{23} = (2 \times 3)^{-1/2} = 0.4082; \quad x_{24} = (2 \times 4)^{-1/2} = 0.3536;$$

$$x_{33} = (3 \times 3)^{-1/2} = 0.3333; \quad x_{34} = (3 \times 4)^{-1/2} = 0.2887;$$

$$x_{44} = (4 \times 4)^{-1/2} = 0.2500;$$

Examples:

$$x_R(G_1) = x_{12} + x_{13} + 3x_{14} + x_{23} + x_{34} = 3.4814$$

$$x_R(G_2) = x_{13} + x_{22} + 2x_{23} + 3x_{33} = 2.8937$$

The molecular connectivity index has been generalized by Kier et al. (1975) so as to include all possible paths[*]

---

[*] A *path* is a walk in which all vertices are distinct. A *walk* of a graph is a sequence, beginning and ending with vertices, in which vertices and edges alternate and each edge is incident with the vertices immediately preceding and following it. The *length* of a path is the number of edges in it.

of length h in the molecular graph but not only the edges {i, j} which are paths of length h = 1;

$$^h\chi_R = \sum_{paths} (v_i v_j \cdots v_{h+1})^{-1/2} \qquad (67)$$

Here v_i, v_j, ... , v_{h+1} stand for the degrees of vertices in the path of length h. h = 1, 2, and 3 is usually dealt with since the interactions between atoms separated by a path of a length larger than three are assumed not to influence molecular properties.

The calculation of the second order molecular connectivity for graph G_1 is shown below:

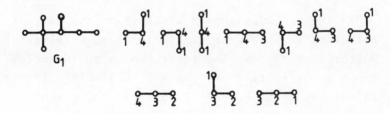

FIG. 12.   Paths of length two in graph G_1. The vertex degrees are labelled to each vertex

$$^2\chi_R(G_1) = {}^3\chi_{141} + {}^3\chi_{143} + \chi_{134} + \chi_{234} +$$

$$+ \chi_{132} + \chi_{123} = 3.6753$$

The molecular connectivity indices have been extensively applied to various correlations with the properties and biological activities of chemical compounds (Kier and Hall, 1976). They also display a high capability of isomer discrimination (Bonchev et al., 1981 b), and reflect some features of molecular branching (Randić, 1975) and cyclicity.

iv. The Hosoya Index, Z(G)

The number of ways in which k edges are chosen from the graph G so that no two of them are adjacent (or otherwise, the *non-adjacent number*), P (G,k), has been used by Hosoya (1971, 1972) in the definition of a new topological index, Z(G):

$$Z(G) = \sum_{\kappa=0}^{[N/2]} P(G,k) \qquad (68)$$

N/2 in the Gauss square brackets is the nearest integer not exceeding the real number in them, N being the total number of vertices. The first two terms in eq (68) are known by definition. P(G,O) = 1, while (P(G,1) = E, the latter being the number of edges in the graph.

The Hosoya index has been applied to the study of molecular branching, to correlations with boiling points and entropies of acyclic and cyclic hydrocarbons, as well as to the coding of chemical structures (Hosoya 1971; 1972; Mizutani et al., 1971; Kawasaki et al., 1971).

Examples:

$P(G_1,2) = 11$:

$P(G_1,3) = 3$;

$$Z(G_1) = 1 + 7 + 11 + 3 = 22$$

Similarly, for graph G_2, one obtains:

$P(G,0) = 1$, $P(G_2,1) = 7$, $P(G_2,2) = 10$, and

$P(G_2,3) = 2$

$$Z(G_2) = 1 + 7 + 10 + 2 = 20$$

v. The Wiener Index, $W(G)$

This is first topological index introduced by Wiener as
far as 1947. Wiener specified his "path number" in an empirical way as the sum of the number of bonds separating
each pair of atoms in hydrocarbon molecules. Hosoya
(1971) pointed out the graph-theoretical background of
the Wiener number and extended its definition so as to

comprise cyclic molecules as well.

The new definition is related to the *distance matrix* of a graph, $D(G)$. This is a square NxN matrix characterizing uniquely any graph G having N vertices. Its entries d_{ij} are termed *distances*. They equal the number of edges in the shortest path connecting any pair of vertices i and j. Thus, $d_{ij} = 1, 2, 3, \ldots, d_{max}$, while diagonal elements $d_{ii} = 0$, by definition.

Examples:

$$D(G_1) = \begin{array}{c|cccccccc|c} & 1 & 2 & 3 & 4 & 5 & 6 & 7 & 8 & d_i \\ \hline 1 & 0 & 1 & 2 & 3 & 4 & 2 & 2 & 3 & 17 \\ 2 & 1 & 0 & 1 & 2 & 3 & 1 & 1 & 2 & 11 \\ 3 & 2 & 1 & 0 & 1 & 2 & 2 & 2 & 1 & 11 \\ 4 & 3 & 2 & 1 & 0 & 1 & 3 & 3 & 2 & 15 \\ 5 & 4 & 3 & 2 & 1 & 0 & 4 & 4 & 3 & 21 \\ 6 & 2 & 1 & 2 & 3 & 4 & 0 & 2 & 3 & 17 \\ 7 & 2 & 1 & 2 & 3 & 4 & 2 & 0 & 3 & 17 \\ 8 & 3 & 2 & 1 & 2 & 3 & 3 & 3 & 0 & 17 \end{array}$$

$$D(G_2) = \begin{array}{c|cccccc|c} & 1 & 2 & 3 & 4 & 5 & 6 & d_i \\ \hline 1 & 0 & 1 & 2 & 3 & 3 & 2 & 11 \\ 2 & 1 & 0 & 1 & 2 & 2 & 1 & 7 \\ 3 & 2 & 1 & 0 & 1 & 2 & 1 & 7 \\ 4 & 3 & 2 & 1 & 0 & 1 & 2 & 9 \\ 5 & 3 & 2 & 2 & 1 & 0 & 1 & 9 \\ 6 & 2 & 1 & 1 & 2 & 1 & 0 & 7 \end{array}$$

In the examples above the distance sums S_i are also given which will be used in the next subsection.

The Wiener index equals the sum of all the entries of the triangular distance submatrix:

$$W(G) = \frac{1}{2} \sum_{ij}^{N} d_{ij} \qquad (69)$$

Making use of eq (64) one obtains: $W(G_1) = 63$ and $W(G_2) = 25$.

At a constant number of atoms the Wiener index has a maximum for the linear n-alkane while it has a minimum for the most compact (most branched and cyclic) structures. Due to this it is a very convenient measure of molecular branching and cyclicity as is shown in a series of papers by Bonchev and Trinajstić (1977, 1978), Bonchev et al. (1978, 1980 a), Mekenyan et al. (1979 a, 1981 a, b). Therefore, it is not surprising that the Wiener index usually provides high correlations with structure dependent properties of chemical compounds. Wiener (1947) and Platt (1947, 1952) reported such correlations with boiling points, heats of formation and vaporization, molecular volume and molecular refraction of alkanes. Rouvray (1975 b, 1976) also found excellent correlations with a number of properties of alkanes, alkenes, and alkynes(melting and boiling points, surface tension, viscosity, density, refraction index), as well as arenes (boiling points). Gas chromatographic retention indices of alkanes, benzenoid hydrocarbons and alkylbenzenes have also been satisfactorily calculated in this way (Bonchev et al., 1979 b; Trinajstić et al., 1979). The Wiener index also found applications to the prediction of the energy gaps and physico-chemical properties of conjugated polymers, as well as to the modelling of crystal growth and crystal vacancies.

vi. The Balaban Distance Connectivity Index, J(G)

Quite recently, Balaban (1982) proposed another topological index based on the distance matrix which seems to
be the most discriminating out of the known single topological indices. It is specified similarly to the Randić
connectivity index (eq 61). The two neighbouring vertices
i and j which form the edge {i,j} are represented in the
new index by means of the so-called *distance sums* of distance degrees, d_i and d_j. The distance sum for each vertex i is the sum of all distance matrix elements in the
ith row or column. The vertex distance sums thus recall
the vertex degrees v_i and v_j used in the formulation of
the Randić index χ_R since the latter are sums of all adjacency matrix elements in the ith (or respectively jth)
row or columm. Evidently, the distance degrees are also
graph-invariants but unlike the vertex degrees they have
a high number of distinct values. It should be mentioned
that the distance degrees have already been used in developing a new graph centre concept (Bonchev et al., 1980 b),
as well as in the study of some general properties of
graphs (Polansky, 1980).
The Balaban distance connectivity index, J, differs also from the Randić connectivity index by the presence of
the normalizing factor A:

$$J = A \sum_{\text{adjacent } i,j} (d_i d_j)^{-1/2} \tag{70}$$

where

$$A = \frac{C}{C + 1} \tag{71}$$

B being the total number of edges (bonds), and C being the total number of cycles in the molecular graph. Clearly, for acyclic graphs A = B.

Examples (See the distance matrices D (G_1) and D (G_2) in the previous subsection):

$$J(G_1) = 7[(d_1 d_2)^{-1/2} + (d_6 d_2)^{-1/2} + (d_7 d_2)^{-1/2} + (d_8 d_3)^{-1/2}$$

$$+ (d_5 d_4)^{-1/2} + (d_4 d_3)^{-1/2} + (d_2 d_3)^{-1/2}] = 7[4(17 \times 11)^{-1/2}$$

$$+ (21 \times 15)^{-1/2} + (15 \times 11)^{-1/2} + (11 \times 11)^{-1/2}] = 3.6233$$

$$J(G_2) = \frac{7}{3}[(d_1 d_2)^{-1/2} + (d_2 d_3)^{-1/2} + (d_2 d_6)^{-1/2} +$$

$$+ (d_3 d_6)^{-1/2} + (d_3 d_4)^{-1/2} + (d_4 d_5)^{-1/2} + (d_6 d_5)^{-1/2}] =$$

$$= \frac{7}{3}[(11 \times 7)^{-1/2} + 3(7 \times 7)^{-1/2} + 2(7 \times 9)^{-1/2} + (9 \times 9)^{-1/2}]$$

$$= 2.1131$$

One might expect the new index to be of great importance for predicting physico-chemical properties and biological activities of molecules.

vii. The Balaban Centric Index, B(G)

Another interesting index has been devised by Balaban (1979) aiming at reflecting the shape of acyclic molecules. It is based on the so-called *pruning partition* of graph vertices N:

$$P_p(G) = N \{a_1, a_2, \ldots, a_k\} \qquad (72)$$

The pruning is a stepwise procedure of removing all terminal vertices (vertices of degree one), a_i is then the
number of vertices deleted upon step i of the procedure
while a_k is the number of central vertices of the graph
determined in the last stage of the pruning (for tree
graphs a_K = 1 or 2). A *central vertex* in a graph is a vertex v_i having the least maximum distance to the other
graph vertices:

$$d_i^{max} = min \quad for \quad i = 1, 2, 3, \ldots, N \qquad (73)$$

The *centre* of a graph is the set of its central vertices
(Harary, 1969).

The centric index of Balaban, $B(G)$, is calculated from
eq (72) by means of a quadratic-type formula:

$$B(G) = \sum_{i=1}^{N} a_i^2 \qquad (74)$$

Example:

$$B(G_1) = 5^2 + 2^2 + 1^2 = 30$$

B(G) has additionally been normalized in order to provide a zero value for the least branched (linear) acyclic
graph, as well as a maximum value B(G) = 1 for the most
branched case (a star-graph). The normalized centric index correlates well with the octane number of acyclic
hydrocarbons (Balaban and Motoc, 1979).

viii. Extended Connectivity Indices

Another way of generalizing the Randić connectivity index
has been proposed by Evans et al. (1978):

$$I_{2AB} = [(^2v_i a_i \times {}^2v_j a_j) b]^{-1/2} \tag{75}$$

and

$$I_{3AB} = [(^3v_i a_i + {}^3v_j a_j) b]^{-1/2} \tag{76}$$

where a_i and a_j are integers describing the atomic types
of i and j; ^2v_i, ^2v_j, ^3v_i, and ^3v_j are the second - and
third-order connectivities of i and j (i.e. the sum of
their second and third neighbours, respectively) and b
is the order of the bond between them. The only atomic types considered by Evans et al. are C, N, and O, their paramet-

ers being arbitrarily chosen as 3, 5, and 7, respectively. Bawden et al. (1981) have tested different types of parametrization for the representation of chemical elements from a much wider range. The two indices have been found to have a high discriminatory power within large sets of different structures, incorporating subsets of structurally similar compounds. The procedure has been implemented in the Pfizer Central Research (U.K.) computerized chemical information system, for interactive registration and structure search.

Another approach to the compact computer representation of a structure has recently been proposed by Freeland et al. (1979). This representation is called the Augmented Connectivity Molecular Formula (ACMF). The latter incorporates strings of symbols accounting for different atom and bond characteristics. The major role in ACMF is played by the so-called augmented connectivity values of the non-hydrogen atoms in the molecule. Some special structural characteristics , such as abnormal valence and mass, isotopes at unknown locations, delocalized charges, tautomer mobile groups etc., are also included in ACMF.The augmented connectivity value is obtained in an iterative procedure similar to the Morgan algorithm (Morgan, 1965), i.e. given the "level n" values for the vertices of the graph, the "level n+1" value is calculated for each vertex by summing the "level n" values over all vertices adjacent to the given vertex. The initial connectivity value of each atom is determined by multiplying the element symbol (specified in a standard table) by the appropriate bond-type number (specified in another standard table) and summing these products over all vertices adjacent to the given vertex.

As an example the 4-(chloromethyl) - 1H-imidazol molecule and its ACMF is shown:

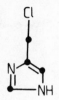

ACMF: C(4864), C(5538),
C(6146), C(8234), Cl(3006),
N(3066), N(4516)

The augmented connectivity molecular formula has been used in a highly efficient algorithm, which is part of the Chemical Abstract Service chemical registry system.

ix. Comparative Tables for the Topological Indices

The values of the topological indices, discussed above, are supplemented in TABLES 8 and 9 for some acyclic (FIG.13) and monocyclic (FIG.14) graphs.

Our brief exposé of the topological indices is thus concluded. It was not meant to review all known indices among which one can find many other interesting numerical descriptors of molecules like those presented by Gordon and Scantlebury (1964), Smolenski (1964), Merrifield and Simmons (1981 a, b), etc. More complete and updated reviews on the subject are presented by Balaban (1980), Trinajstić (1982), Sablić and Trinajstić (1982), and Balaban et al. (1982).

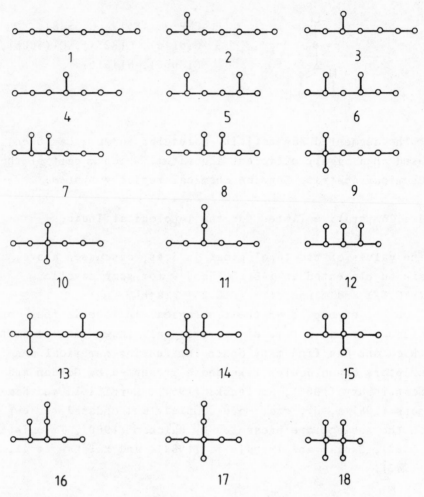

FIG.13. Graphs of acyclic hydrocarbons with eight carbon atoms

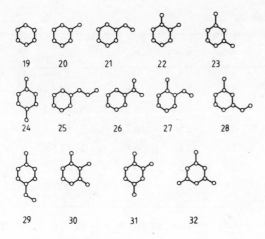

FIG.14.   Graphs of monocyclic hydrocarbons (alkylbenzenes) with six to none carbon atoms

TABLE 8.   Six topological indices for acyclic hydrocarbons having eight carbon atoms (the Zagreb index, M; the Balaban centric index, B; the Hosoya index, Z; the Wiener index, W; the Randić connectivity index, x_R; and the Balaban distance connectivity index, J)

Graphs[a]	M	B	Z	W	x_R	J
1	26	16	34	84	3.9142	2.5301
2	28	18	29	79	3.7701	2.7160
3	28	18	31	76	3.8081	2.8621
4	28	18	30	75	3.8081	2.9196
5	30	24	25	74	3.6259	2.9278
6	30	24	26	71	3.6639	3.0988
7	30	24	27	70	3.6807	3.1708
8	30	24	29	68	3.7188	3.2925
9	32	24	23	71	3.5607	3.1118
10	32	24	25	67	3.6213	3.3734
11	28	22	32	72	3.8510	3.0744
12	32	30	24	65	3.5534	3.4973

Table 8 (Continued)

Graphs[a]	M	B	Z	W	x_R	J
13	34	30	19	66	3.4165	3.3889
14	34	30	22	63	3.4814	3.6233
15	34	30	23	62	3.5040	3.7083
16	30	26	28	67	3.7188	3.3549
17	32	26	28	64	3.6819	3.5832
18	38	40	17	58	3.2500	4.0204

a) Numbers correspond to the graphs given in FIG. 13

TABLE 9. Five topological indices[a] for monocyclic hydrocarbons (alkylbenzenes) having six to nine carbon atoms

Graphs[b]	M	Z	W	x_R	J
19	24	18	27	3.0000	2.0000
20	30	26	42	3.3940	2.1229
21	34	44	64	3.9320	2.1250
22	36	39	60	3.8047	2.2794
23	36	37	61	3.7880	2.2307
24	36	38	62	3.7880	2.1924
25	38	70	94	4.4320	2.0779
26	40	62	88	4.3047	2.2284
27	40	65	86	4.3427	2.2973
28	40	63	88	4.3260	2.2317
29	40	64	90	4.3260	2.1804
30	42	58	82	4.2154	2.4017
31	42	56	84	4.1987	2.4072
32	42	52	84	4.1820	2.3408

a) The same indices as in Table 8
b) The numbers correspond to the graphs given in FIG.14

B. THE FIRST TOPOLOGICAL INFORMATION INDICES

i. Orbital Information Indices

The first theoretic-information considerations of chemical
systems have been closely connected with the information
aspects of basic biological phenomena. From the point of
view of information theory an organism must have a very
large information content in order to perform its vital
functions. Thus, a lower limit of complexity exists even
for the simplest organisms. The numerical estimates of
complexity of living organisms, based on the different
kinds of atoms in them, are not large enough, and molecules have to be taken into consideration. As pointed out by
Quastler (1953) and Gamov (1954a,b) the situation is similar to the formation of a very large number of distinct
words from a limited number of letters in the alphabet,
hence, the analogy between the information content of a
written text and that of some organic molecules. The
chemically different atoms or their chemically different
aggregates (molecules) play the role of letters and words.
 As shown by Rashevsky (1955) a molecule, composed of
chemically indistinguishable units (atoms), can also have
large information content. The atoms in this case are
different through their relationships to each other.
Rashevsky made use of the topological differentiation of
atoms in molecules as expressed by the molecular graphs.
He introduced the *topological information content* of a
molecule:

$$\bar{I}_{top} = - \sum_{i=1}^{k} \frac{N_i}{N} \, lb \, \frac{N_i}{N}, \text{ bits per atom} \tag{77}$$

where N_1 stands for the number of topologically equivalent atoms in the ith subset of atoms.

As a first approximation all the atoms having the same valence are considered equivalent. An additional topological distinction can be introduced taking into account the second, third, etc., neighbours. For instance there are three vertices of degree two (1,4,5) and two others of degree three (2,3) in the structure:

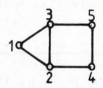

Taking into account the second neighbours, one discriminates between vertex 1 and the vertices 4 and 5, leaving the latter indistinguishable. (Thus the two second neighbours of vertex 1, are of degree two $v_4 = v_5 = 2$, while the two second neighbours of vertices 4 and 5 are of degree 2 and 3, respectively: $v_1 = 2$, $v_3(v_2) = 3$).

The topological equivalence of two atoms in a molecule has been expressed above as equivalence of two vertices, s and t, in the graph so that for each ith neighbouring vertex (i = 1, 2, 3, ..., k) of vertex s there exists an ith neighbouring vertex of the same degree for vertex t. This condition is, however, necessary but not sufficient, as can be seen from example G_3.

In G_3, vertices 1 and 1' are topologically non-equivalent, irrespective of the equivalent vertex degree sets S and S'-, since they belong to 5- and 6- membered cycles, respectively.

Trucco (1957 a, b) based the definition of topological information on the *graph orbits*. Each graph belongs to a

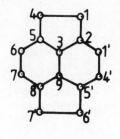

G_3

ver- tices	sets of respective neighbouring vertices of the same degrees
1	S {2,3,4,5,6,7,8,9}
1'	S'{2,3,4',5',6',7',8,9}

certain automorphism group where automorphisms are vertex permutations preserving adjacency of the graph. It is those atoms, which belong to the same orbit of the automorphism group, that are regarded as topologically equivalent, i.e. the vertices which can interchange preserving the adjacency of the graph.

Hence, N_i in eq (77) should be understood as the cardinality of the ith orbit of the graph.In graph G_3 vertices 1 and 1' are not topologically equivalent because they belong to different orbits {1,4,6',7'} and {1',4',6,7}, respectively.

Due to the existence of other information indices for graphs, each of them reflecting some topological feature of the structures, the term "topological information" is more general than initially supposed. For this reason the topological information of Rashevsky and Trucco will be denoted further as *information on the orbits of the vertex automorphism group of a graph* (or shortly, orbital information), $^V\bar{I}_{ORB}$.

Trucco (1956 a,b) also pointed out that the same approach could be applied to graph edges, proceeding from the edge automorphism group and the edge orbits of the graph. Therefore, for each graph distinction should be made between *vertex orbital information*, $^V I_{ORB}$, and *edge orbital information,* $^E I_{ORB}$.

Examples:

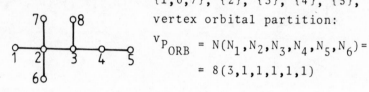

vertex orbits:
{1,6,7}, {2}, {3}, {4}, {5}, {8}
vertex orbital partition:
$$^v P_{ORB} = N(N_1,N_2,N_3,N_4,N_5,N_6) =$$
$$= 8(3,1,1,1,1,1)$$

G_1

$$^v \bar{I}_{ORB} (G_1) = - 5x \frac{1}{8} \text{ lb } \frac{1}{8} - \frac{3}{8} \text{ lb } \frac{3}{8} = 2.4056 \text{ bits}$$

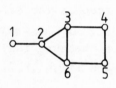

vertex orbits: {1}, {2}, {3,6},{4,5}
vertex orbital partition:
$$^v P_{ORB} = 6(1,1,2,2)$$

G_2

$$^v \bar{I}_{ORB}(G_2) = - 2x \frac{1}{6} \text{ lb } \frac{1}{6} - 2x \frac{2}{6} \text{ lb } \frac{2}{6} = 1.9183 \text{ bits}$$

Examples:

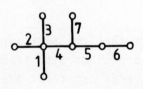

edge orbits:
{1,2,3}, {4}, {5}, {6}, {7}
edge orbital partition:
$$^E P_{ORB} = E(E_1,E_2,E_3,E_4,E_5) =$$
$$= 7(3,1,1,1,1)$$

G_1

$$^E\bar{I}_{ORB}\ (G_1) = -\ 4x\ \frac{1}{7}\ lb\ \frac{1}{7} - \frac{3}{7}\ lb\ \frac{3}{7} = 2.1281\ bits$$

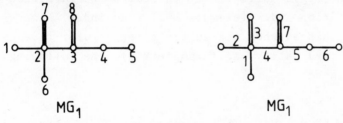

edge orbits:

{1}, {2,3}, {4}, {5,6}   {7}

edge orbital partition:

$$^E P_{ORB} = 7(1,1,1,2,2)$$

G_2

$$^E\bar{I}_{ORB}\ (G_2) = -\ 3x\ \frac{1}{7}\ lb\ \frac{1}{7} - 2x\ \frac{2}{7}\ lb\ \frac{2}{7} = 2.2359\ bits$$

The vertex and edge orbital information indices can also be specified for multigraphs, MG.

Examples:

vertex orbits: {1,2}, {3}, {4}, {5}, {6}, {7}, {8}

$$^V P_{ORB}\ (MG_1) = 8(2,1,1,1,1,1,1)$$

$$^V\bar{I}_{ORB}\ (MG_1) = -\ 6x\ \frac{1}{8}\ lb\ \frac{1}{8} - \frac{2}{8}\ lb\ \frac{2}{8} = 2.7500\ bits$$

edge orbits: {1,2}, {3}, {4}, {5}, {6}, {7}

$$^E P_{ORB}\ (MG_1) = 7(2,1,1,1,1,1)$$

$$^E\bar{I}_{ORB}\ (MG_1) = -\ 5x\ \frac{1}{7}\ lb\ \frac{1}{7} - \frac{2}{7}\ lb\ \frac{2}{7} = 2.5216\ bits$$

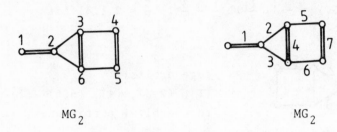

$$MG_2 \qquad\qquad MG_2$$

vertex orbits: $\{1\}$, $\{2\}$, $\{3,6\}$, $\{4,5\}$
edge orbits: $\{1\}$, $\{2,3\}$, $\{4\}$, $\{5,6\}$, $\{7\}$

$${}^{V}\bar{I}_{ORB}\,(MG_2) = {}^{V}\bar{I}_{ORB}(G_2)$$

$${}^{E}\bar{I}_{ORB}\,(MG_2) = {}^{E}\bar{I}_{ORB}(G_2)$$

Comparing the orbital information indices of graphs and multigraphs, evidently, the multiplicity of edges as a rule introduces additional distinction between the graph edges; this multiplicity increases the magnitude of the information index (example MG_1). The index, however, remains unchanged when the multiple bonds refer to edge orbits of cardinality one (example MG_2). The following relations hold:

$$
{}^{V}\bar{I}_{ORB}\,(MG) \geqslant {}^{V}\bar{I}_{ORB}\,(G) \qquad\qquad (78)
$$

$$
{}^{E}\bar{I}_{ORB}\,(MG) \geqslant {}^{E}\bar{I}_{ORB}(G) \qquad\qquad (79)
$$

Finally, both vertex and edge orbital information indices can be defined for directed graphs, DG (Rashevsky, 1955; Mowshovitz, 1968 b). The edges (arcs) in these graphs may additionally differ in their orientations. The

vertices can similarly differ being initial or end points
of an arc.

Hence relations analogous to (73) and (74) are valid:

$$^{E}\overline{I}_{ORB} \ (DG) \geqslant \ ^{E}\overline{I}_{ORB} \ (G) \qquad\qquad (80)$$

$$^{V}\overline{I}_{ORB} \ (DG) \geqslant \ ^{V}\overline{I}_{ORB} \ (G) \qquad\qquad (81)$$

Example:

$$DG_2$$

vertex orbits: {1}, {2}, {3}, {4}, {5}, {6}
edge orbits: {1}, {2}, {3}, {4}, {5}, {6}, {7}

$$^{V}P_{ORB} = 6(1,1,1,1,1,1); \qquad ^{E}P_{ORB} = 7(1,1,1,1,1,1,1)$$

$$^{V}\overline{I}_{ORB} \ (DG_2) = 1b \ 6 = 2\text{-}5850 \qquad ^{V}\overline{I}_{ORB} \ (G_2) = 1.9183$$

$$^{E}\overline{I}_{ORB} \ (DG_2) = 1b \ 7 = 2.8074 \qquad ^{E}\overline{I}_{ORB} \ (G_2) = 2.2359$$

Similarly to Morowitz (1955), Rashevsky (1955) also
determined the *total information content* of a molecule as
a sum of a chemical and a structural term, by adding the
information on atomic composition (eq 54) to his (vertex)
orbital information:

$$\bar{I}_{RASH} = \bar{I}_{AC} + \bar{I}_{ORB} \tag{82}$$

Making use of the orbital information index, Karreman (1955) studied for the first time the *information balance of chemical reactions* and found it to be positive as well as negative, depending essentially on the structure of the reagents and reaction products.

Examples:

a)

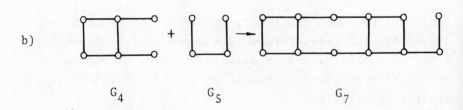

$$G_4 \qquad\qquad G_5 \qquad\qquad\qquad G_6$$

$${}^{v}\bar{I}_{ORB} (G_4) = -3x \frac{2}{6} lb \frac{2}{6} = 1.585 \text{ bits}$$

$${}^{v}\bar{I}_{ORB} (G_5) = -2x \frac{2}{4} lb \frac{2}{4} = 1 \text{ bit}$$

$${}^{v}\bar{I}_{ORB} (G_6) = -5x \frac{2}{10} lb \frac{2}{10} = 2.322 \text{ bits}$$

$$\Delta I (a) = I(G_6) - [I(G_4) + I(G_5)] = -0,263 \text{ bits}$$

b)

$$G_4 \qquad\qquad G_5 \qquad\qquad\qquad G_7$$

$$^{v}\bar{I}_{ORB} \ (G_7) \ = \ - \ 10x \ \frac{1}{10} \ lb \ \frac{1}{10} \ = \ 3.322 \ bits$$

$$\Delta I \ (b) \ = \ I \ (G_7) \ - \ [\ I(G_4) \ + \ I(G_5)] \ = \ + \ 0.737 \ bits$$

As seen from the above examples, the change in the orbital information index may be negative or positive, depending on the different relative orientation of the same two reagents during the reaction, since reaction products of different topology are obtained.

Consider now another reaction with same molecule represented by graph G_5:

$$G_8 \qquad\qquad G_5 \qquad\qquad\qquad G_9$$

$$^{v}\bar{I}_{ORB} \ (G_8) \ = \ - \ 1 \ lb \ 1 \ = \ 0$$

$$^{v}\bar{I}_{ORB} \ (G_9) \ = \ - \ \frac{2}{10} \ lb \ \frac{2}{10} \ - \ 2x \ \frac{4}{10} \ lb \ \frac{4}{10} \ = \ 1.522 \ bits$$

$$\Delta I(c) \ =I(G_9) \ - \ [\ I(G_8) \ + \ I(G_5)] \ = \ 0.522 \ bits$$

The difference between $\Delta I(c)$ and $\Delta I(a)$,

$$\Delta I(c) \ - \ \Delta I(a) \ = \ 0.785 \ bits$$

is regarded by Karreman (1955) as a quantitative measure of the difference in structural specificity of the two molecules G_4 and G_8 for the molecule G_5.

It has already been mentioned in the beginning of this subsection that Rashevsky (1955) brought into view the biological importance of the topological information. He proceeded from the fact, that the basic vital functions of an organism (selection of food, breaking up of the food molecules into appropriate parts, selection of those parts, and their assimilation) are to a great extent determined by its information content, and therefore, by the information content of the constituent organic molecules. A close relation between topology and life was thus found on a molecular level.

In a later work Rashevsky (1960) proposed a general method and presented numerical calculations for the minimum amount of topological (orbital) information which was necessary for the self-reproduction of an organism. The fundamental problem of the possibility of life's spontaneous generation on Earth was explored on this basis. It was found that a system containing less than 150 bits of specific information could be generated by means of a pure chance process. This is much less than is usually ascribed to the information content of genes or DNA molecules. Thus, even under the most favourable conditions, the chances of life ever having been formed not only on Earth but even in the whole universe were estimated as vanishingly small. It was concluded that dynamic factors, which may reduce tremendously the information needed, must play a major role in the genesis of life on Earth.

In a series of papers Mowshowitz (1968a,b,c) developed further the Rashevsky method applying it to the relative complexity of undirected and directed graphs. The topological information is an appropriate measure of complexity, since it depends both on the number of equivalence classes of the graph vertices and edges and on their respective cardinalities. The change in this structural index upon various graph operations (complement, sum, join, cartes-

ian product and composition) might be applied to the
topological-information modelling of chemical reactions.
Exploring this change in detail, Mowshowitz found conditions for semi-additivity of information, which are an
important contribution to the problem of the information
balance of chemical reactions. Conditions have been also
found for a graph to have a prescribed information content,
as well as for two graphs to have equal information content

ii. The Chromatic Information Index

A *colouring* of a graph G is an assignment of minimal number
of colours to the vertices of G such that no two adjacent
vertices have the same colour. A decomposition $V(V_1, V_2, \ldots,$
$V_k)$ of the set $V(G)$ of the graph vertices is said to be a
chromatic decomposition of G, $P_{CHR}(G)$, if for any pair of
vertices $x, y \in V_i$ the edge $\{x, y\}$ does not belong to the
set of graph edges $E(G)$. The sets V_i are called *colour
classes*. The *chromatic number* $k(G)$ is the smallest number
k for which G has a chromatic decomposition with k colour
classes.

A second type of information index for graphs, called
chromatic information content, $\bar{I}_{chr}$, has been introduced
by Mowshowitz (1968 d). As, in general, there is no unique chromatic decomposition of a graph, it is difficult
to specify an information measure which should reflect in
a unique way the chromatic structure of a graph. For this
reason I_{chr} is defined by eq (6) and (7) as the minimum
information over all finite probability schemes constructed from vertex chromatic decompositions having a rank
equal to the chromatic number of a graph:

94

$$
{}^{v}\bar{I}_{chr}\ (G) = \min_{\hat{V}}\{-\sum_{i=1}^{k(G)} \frac{N_i(\hat{V})}{N}\ \text{lb}\ \frac{N_i(\hat{V})}{N}\} \tag{83}
$$

where $\hat{V}$ is an arbitrary chromatic decomposition of a graph G, and k is the chromatic number of G.

Examples:

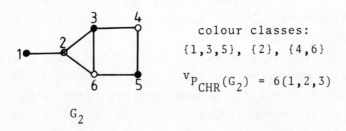

colour classes:
{1,3,5,6,7}, {2,4,8}

$${}^{v}P_{CHR}(G_1) = 8(5,3)$$

G_1

$$
{}^{v}\bar{I}_{chr}(G_1) = -\frac{5}{8}\ \text{lb}\ \frac{5}{8} - \frac{3}{8}\ \text{lb}\ \frac{3}{8} = 0.9544\ \text{bits}
$$

colour classes:
{1,3,5}, {2}, {4,6}

$${}^{v}P_{CHR}(G_2) = 6(1,2,3)$$

G_2

$$
{}^{v}\bar{I}_{CHR}(G_2) = -\frac{1}{6}\ \text{lb}\ \frac{1}{6} - \frac{2}{6}\ \text{lb}\ \frac{2}{6} - \frac{3}{6}\ \text{lb}\ \frac{3}{6} = 1.4591\ \text{bits}
$$

In G_1 and G_2, there is a unique chromatic decomposition. A graph with different chromatic decompositions is shown in FIG.15.

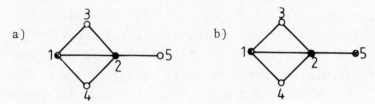

a) b)

FIG.15.   Graph with two different vertex chromatic decompositions

colour classes   chromatic decompositions

$\hat{v}_a$: {1}, {2}, {3,4,5} $^vP_{CHR}(G_a) = 5(1,1,3)$

$\hat{v}_b$: {1,5}, {2}, {3,4} $^vP_{CHR}(G_b) = 5(1,2,2)$

$$^v\bar{I}_{CHR}(G_a) = - 2x \frac{1}{5} \text{ lb } \frac{1}{5} - \frac{3}{5} \text{ lb } \frac{3}{5} = 1,3710 \text{ bits}$$

$$^v\bar{I}_{CHR}(G_b) = - \frac{1}{5} \text{ lb } \frac{1}{5} - 2x \frac{2}{5} \text{ lb } \frac{2}{5} = 1,5219 \text{ bits}$$

Clearly, $\hat{v}_{min} = \hat{v}_a$,   and $^v\bar{I}_{CHR}(G) = {}^v\bar{I}_{CHR}(G_a)$.

Mowshowitz (1968 d) proved that:

$$^v\bar{I}_{CHR}(G) \leqslant \text{lb } k(G) \tag{84}$$

and $$^v\bar{I}_{CHR}(G) \leqslant \text{lb } (v_{max} + 1) \tag{85}$$

The latter follows from the fact that the chromatic number of a graph $k(G) \leqslant (v_{max} + 1)$, v_{max} being the maximum vertex degree in G.

Comparing the two information indices defined for

96

graphs, the vertex orbital and chromatic information
indices, Mowshowits (1968 d) showed that there is only
little coincidence between them. Thus for monocyclic
structures of any size the orbital information is always
zero because all the graph vertices belong to the same
orbit of the automorphism group. The chromatic information
of such a graph is always non-zero, the chromatic number
being two or three for even- and odd- membered cycles,
respectively. For acyclic graphs the chromatic number is
two, hence:

$$^v\bar{I}_{CHR}(\text{acyclic graph}) \leqslant 1 \tag{86}$$

For acyclic graphs with more than six vertices, however,
there always exists a graph G without any automorphism
other than identity, whose orbital information content
$^v\bar{I}_{ORB}(G) = \text{lb } N$ is much larger than one. Still one can
find many cases of graphs having the same orbital and
chromatic information index, e.g. FIG.16, 17

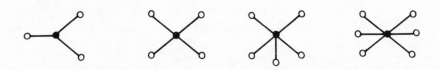

FIG.16. Star-graphs having the same chromatic and
orbital information contents

$$^v\bar{I}_{ORB}(\text{Star}) = {}^v\bar{I}_{CHR}(\text{Star}) = -\frac{1}{N} \text{ lb } \frac{1}{N} - \frac{N-1}{N} \text{ lb } \frac{N-1}{N} \tag{87}$$

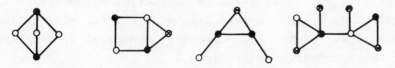

FIG.17. Cyclic graphs having the same chromatic and orbital information contents

A simple result is obtained for even monocyclic and chain hydrocarbons whose C- graphs are bicolourable (FIG.18):

$$^VP_{CHR} = N(N/2, N/2)$$

$$^V\bar{I}_{CHR} \text{ (even cycles and chains)} = 1 \text{ bit} \tag{88}$$

FIG.18. Bicolourable graphs of even alternant (monocyclic and chain) hydrocarbons each carbon atom in which carries exactly 1 bit vertex chromatic information

The edge chromatic information index, $^E\bar{I}_{CHR}(G)$ is defined (Bonchev and Mekenyan, 1982 a) analogously to the vertex chromatic index replacing in all definitions "vertex" with "edge":

$$^E\bar{I}_{CHR}(G) = \min_{\hat{E}} \{-\sum_{i=1}^{k_E(G)} \frac{B_i(\hat{E})}{B} \text{ lb } \frac{B_i(\hat{E})}{B}\} \tag{89}$$

where $\hat{E}$ is an arbitrary edge chromatic decomposition of a graph, G, B_i is the cardinality of the ith edge colour class, $k_E(G)$ is the edge chromatic number of G, and B is the total number of edges in G.

Evidently,

$$k_E(G) \geqslant v_{max} \qquad (90)$$

which in general makes the number of edge colour clases larger than those of vertex colour classes. Acyclic graphs are the simplest illustration since they are vertex bicolourable while having vertices of degree 3 and 4 they are three- and four- edge colourable.

Examples: edge chromatic decomposition:

$${}^E P_{CHR}(G_1) = 7(3,2,1,1)$$

$${}^E \bar{I}_{CHR}(G_1) = 1.8424 \text{ bits}$$

$${}^E P_{CHR}(G_2) = 7(3,2,2)$$

$${}^E \bar{I}_{CHR}(G_2) = 1.5567 \text{ bits}$$

Unlike eq (87), star- graphs being B - colourable have a maximim edge chromatic information while their edge orbital information is zero:

$${}^E \bar{I}_{CHR}(\text{Star}) = \text{lb } B = \text{max} \qquad (91)$$

For even cycles and chains the maximum vertex and edge degrees equal two, hence eq (88) holds for both vertex and edge chromatic information indices.

Multigraphs and digraphs have in principle the same chromatic information content as their respective parent graphs since the presence of double or triple edges, as well as the presence of directed edges (arcs) does not change the graph colouring. The edge chromatic information, however, could be made distinct for digraphs and multigraphs if the definition of edge colouring is supplemented by the prohibition for two edge to have the same colour if they are of different multiplicity or have different directions.

Examples:

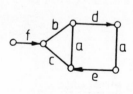

MG$_2'$

In MG$_2'$ the terminal edge is single and cannot be coloured by the same colour as the two double edges a. Hence

$$^E P_{CHR}(MG_2') = 7(2,2,2,1)$$

$$^E \bar{I}_{CHR}(MG_2') = 1.9502 \text{ bits} \neq {}^E\bar{I}_{CHR}(G_2)$$

DG$_2$

In DG$_2$ edge f, being directed, cannot be coloured by the same colour as the two undirected edges a. Moreover, the comparison with graph G$_2$ shows that the pairs of edges b and c in G$_2$ have to be coloured in DG$_2$ by four different colours because d and e are arcs and they are oriented in the opposite directions (d originates from, while e terminates in, a vertex of degree 3).

$$E_P(DG_2) = 7(2,1,1,1,1,1)$$

$$^E\bar{I}_{CHR}(DG_2) = 2.5216 \neq \bar{I}_{CHR}(G_2)$$

Due to the fact that the presence of multiple edges in multigraphs, as well as the presence of oriented edges in directed graphs can split some edge colour classes into two or more new classes or, in the bound case, to preserve the same classes, inequalities (92) and (93) follow:

$$^E\bar{I}_{CHR}(MG) \geqslant {}^E\bar{I}_{CHR}(G) \tag{92}$$

$$^E\bar{I}_{CHR}(DG) \geqslant {}^E I_{CHR}(G) \tag{93}$$

The chromatic information indices can be applied to different problems in chemistry and primarily to the alternant hydrocarbons which display interesting properties. They can be related for instance to the number of Kekule structures of organic compounds. Thus, Bonchev and Mekenyan (1982 a) showed that molecules whose vertices and edges in the graphs carry 1 bit each of chromatic information have always Kekule structures.

C. INFORMATION INDICES BASED ON THE GRAPH ADJACENCY MATRIX

i. Some Definitions

The adjacency matrix of a graph G, $A(G)$, is the most

commonly used graph-theoretical matrix. It is a square matrix whose rows and columns are labelled by the vertices of the graph, the elements being defined by

$a_{ii} = 0$; $a_{ij} = 1$ if vertices i and j are adjacent (linked by an edge); $a_{ij} = 0$ otherwise.

The vertex degree v_i of each vertex i in the graph can be obtained from the adjacency matrix by summation over all entries in the ith row or column:

$$v_i = \sum_{j=1}^{N} a_{ij} \qquad (94)$$

The sum over all entries of $A(G)$ may be called (Bonchev et al., 1980 b) the *total adjacency* of G, $A(G)$:

$$A(G) = \sum_{i=1}^{N} \sum_{j=1}^{N} a_{ij} = \sum_{i=1}^{N} v_i \qquad (95)$$

The total adjacency of G is twice the number of edges in it, B (e.g. Harary, 1969)

$$A(G) = 2B \qquad (96)$$

The quantity B has been used by Burton in studies on clusters though without referring to its graph theoretical origin.

Examples:

	V_1	V_2	V_3	V_4	V_5	V_6	V_7	V_8	v_i
$-V_1$	0	1	0	0	0	0	0	0	1
V_2	1	0	1	0	0	1	1	0	4
$A(G_1) = V_3$	0	1	0	1	0	0	0	1	3
V_4	0	0	1	0	1	0	0	0	2
V_5	0	0	0	1	0	0	0	0	1
V_6	0	1	0	0	0	0	0	0	1
V_7	0	1	0	0	0	0	0	0	1
V_8	0	0	1	0	0	0	0	0	1

	V_1	V_2	V_3	V_4	V_5	V_6	v_i
V_1	0	1	0	0	0	0	1
V_2	1	0	1	0	0	1	3
$A(G_2) = V_3$	0	1	0	1	0	1	3
V_4	0	0	1	0	1	0	2
V_5	0	0	0	1	0	1	2
V_6	0	1	1	0	1	0	3

$$A(G_1) = 14; \quad A(G_2) = 14$$

ii. Information Indices on the Graph's Adjacency Matrix Elements

The set of adjacency matrix elements consists of two subsets: one of zeros and the other of units. The cardinalities of the two subsets can be denoted by N_o and N_1, respectively. Proceeding from the criterion chosen for grouping of a_{ij} into these two subsets, i.e. from the

equality of their magnitudes, an *information index on the equality of (vertex) adjacency matrix elements*, $^{v}\bar{I}^{E}_{adj}$, has been defined by Bonchev and Trinajstic (1977):
(vertex) adjacency matrix elements partition:

$$^{v}P^{E}_{adj} = N^{2}\{N_{1}, N_{o}\}$$

$$^{v}\bar{I}^{E}_{adg} (G) = - \frac{N_{1}}{N^{2}} \text{lb} \frac{N_{1}}{N^{2}} - \frac{N_{o}}{N^{2}} \text{lb} \frac{N_{o}}{N^{2}} =$$

$$= - \frac{2B}{N^{2}} \text{lb} \frac{2B}{N^{2}} - \frac{(N^{2}-2B)}{N^{2}} \text{lb} \frac{(N^{2}-2B)}{N^{2}} \tag{97}$$

The following equations have been obtained for the total information index, $^{v}I^{E}_{adj}$, of acyclic and monocyclic graphs:
for acyclic graphs:

$$^{v}I^{E}_{adj} = N^{2} \text{ lb } N^{2} - 2(N-1) \text{ lb}(N-1) - [(N-1)^{2} + 1] \text{ lb}$$
$$[(N-1)^{2} + 1] \tag{98}$$

for monocyclic graphs:

$$^{v}I^{E}_{adj} = N^{2} \text{ lb } N^{2} - N(N-2) \text{ lb } (N-2) - 2N \tag{99}$$

Choosing another criterion for grouping of the adjacency matrix elements, their *magnitude*, another information index can be specified, $^{v}I^{M}_{adj}$ *(information on the magnitude of the adjacency matrix elements)*. For graphs without multi-edges or directed edges this is a rather trivial quantita-

tive measure since the zeros do not contribute to
the total adjacency A and the information index is a
logarithmic function of the number of graph edges:
adjacency magnitude partition:

$$^{v}P^{M}_{adj} \; (G) = A\{1,1,1,\ldots \; , 1\} = A\{2Bx1\}$$

$$^{v}\bar{I}^{M}_{adj} = - 2B \; x \; \frac{1}{A} \; lb \; \frac{1}{A} = lb \; A = lb \; 2B = 1 + lb \; B \qquad (100)$$

This index could be of use when comparing multigraphs
and directed graphs with their parent graphs (vide infra).
Examples:

$$^{v}P^{E}_{adj} \; (G_1) = 8^2 \; \{14, \; 50\}$$

$$^{v}\bar{I}^{E}_{adj} \; (G_1) = - \frac{14}{64} \; lb \; \frac{14}{64} - \frac{50}{64} \; lb \; \frac{50}{64} = 0.7579$$
$$\text{bits}$$

$$^{v}P^{M}_{adj} \; (G_1) = 14\{14x1, \; 50x0\}$$

$$^{v}\bar{I}^{M}_{adj} \; (G_1) = lb \; 14 = 3.8074 \; \text{bits}$$

$$^{v}P^{E}_{adj} \; (G_2) = 6^2\{14,22\}$$

$$^{v}\bar{I}_{adj} \; (G_2) = - \frac{14}{36} \; lb \; \frac{14}{36} - \frac{22}{36} \; lb \; \frac{22}{36} = 0.9641$$
$$\text{bits}$$

$$^{v}P^{M}_{adj} (G_2) = 14 \{14x1, 50x0\}$$

$$^{v}\bar{I}^{M}_{adj} (G_2) = lb\ 14 = 3.3074\ bits$$

iii. Information Indices on the Vertex Degrees of a
Graph

Another pair of information indices can be formulated
proceeding from the degrees of graph vertices. An "equality" type and a "magnitude" type of partitions are of use:
vertex degree partition:

$$^{v}P^{E}_{adj,\ deg} (G) = N \{N_{v_1},\ N_{v_2},\ \ldots,\ N_{v_k}\}$$

vertex degree magnitude partition:

$$^{v}P^{M}_{adj,\ deg} (G) = 2B\ (v_1,\ v_2,\ \ldots,\ v_N)$$

where N_{v_i} is the cardinality of the subset of vertices
having the same degree v_i and k is the number of different degrees realized in G

*The information indices on the vertex degree equality
and magnitude of a graph,* $^{v}\bar{I}^{E}_{adj,deg} (G)$ *and* $^{v}\bar{I}^{E}_{adj,deg}(G)$,
have been defined on this basis (Bonchev and Mekenyan,
1982 a):

$$^{v}\bar{I}^{E}_{adj,\ deg} (G) = -\sum_{i=1}^{k} \frac{N_{v_i}}{N}\ lb\ \frac{N_{v_i}}{N} \tag{101}$$

$$^{v}\overline{I}^{M}_{adj,deg} (G) = - \sum_{i=1}^{max} \frac{v_i}{2B} \, 1b \, \frac{v_i}{2B} \tag{102}$$

Examples:

$$^{v}P^{E}_{adj,deg} (G_1) = 8(5,1,1,1); \quad ^{v}\overline{I}^{E}_{adj,deg} (G_1) = 1.5488 \, bits$$

$$^{v}P^{M}_{adj,deg} (G_1) = 14(4,3,2,5x1); \quad ^{v}\overline{I}^{M}_{adj,deg} (G_1) = 2.7535 \\ bits$$

$$^{v}P^{E}_{adj,deg} (G_2) = 6(3,2,1); \quad ^{v}\overline{I}^{E}_{adj,deg} (G_2) = 1.4591 \, bits$$

$$^{v}P^{M}_{adj,deg} (G_2) = 14(3,3,3,2,2,1);$$

$$^{v}\overline{I}^{M}_{adj,deg} (G_2) = 2.5027 \, bits$$

The information indices defined by eq (101, 102) can for some simple classes of graphs be expressed as a function of the number of vertices N, or bonds B, (Bonchev and Mekenyan, 1982 a): monocycles: $v_i = 2 = const$,

$$^{v}\overline{I}^{E}_{adj,deg} (G) = 0 \tag{103}$$

$$^{v}\overline{I}^{M}_{adj,deg} (G) = 1b \, B = \frac{1}{2} \, ^{v}\overline{I}^{M}_{adj} (G) \tag{104}$$

chains (e.g. normal alkane skeletal graphs):

$$v_1 = 1, \ v_2 = 2; \ N_{v_i} = 2, \ N_{v_2} = N - 2$$

$$^{v}\overline{I}^{E}_{adj,deg}\ (G) = -\frac{2}{N}\ lb\ \frac{2}{N} - \frac{N-2}{N}\ lb\ \frac{N-2}{N} \tag{105}$$

$$^{v}\overline{I}^{M}_{adj,deg}\ (G) = lb(N-1) + \frac{1}{N-1} = lb\ B + \frac{1}{B} \tag{106}$$

iv. Information Indices Based on the Edge Adjacency Matrix

The information theoretic considerations, made in sub-section 3, dealt with the vertex adjacency matrix of a graph. They can be repeated for the *edge adjacency matrix* of a graph, $A_E(G)$, the rows and columns of which are lab-elled by the edges of the graph. The matrix elements of $A_E(G)$ are $a_{ij} = 1$ for i,j being neighbouring (edges incid-ent to a common vertex), and $a'_{ij} = 0$ otherwise. The deg-ree of edge i, e_i, is similarly defined as the sum over all entries of ith row or column of $A_E(G)$.

$$e_i = \sum_{j=1}^{B} a'_{ij} \tag{107}$$

Examples:

	E_1	E_2	E_3	E_4	E_5	E_6	E_7	e_i
E_1	0	1	1	1	0	0	0	3
E_2	1	0	1	1	0	0	0	3
E_3	1	1	0	1	0	0	0	3
E_4	1	1	1	0	1	0	1	5
E_5	0	0	0	1	0	1	1	3
E_6	0	0	0	0	1	0	0	1
E_7	0	0	0	1	1	0	0	2

$$A_E(G_1) =$$

	E_1	E_2	E_3	E_4	E_5	E_6	E_7	e_i
E_1	0	1	1	0	0	0	0	2
E_2	1	0	1	1	1	0	0	4
E_3	1	1	0	1	0	1	0	4
E_4	0	1	1	0	1	1	0	4
E_5	0	1	0	1	0	0	1	3
E_6	0	0	1	1	0	0	1	3
E_7	0	0	0	0	1	1	0	2

$$A_E(G_2) =$$

The *total edge adjacency* of a graph, A_E (G), is similarly defined as

$$A_E(G) = \sum_{i=1}^{B} \sum_{j=1}^{B} a'_{ij} = \sum_{i=1} e_1 \tag{108}$$

Thus, $A_E(G_1) = 20$, and $A_E(G_2) = 22$.

The total edge adjacency of a graph had already been used as a topological index by Platt (1947, 1952) who had defined it as the first neighbour sum. The Platt number has been applied to the prediction of some properties of alkanes such as boiling points, heats of formation, heats of vaporization, molar volumes, etc.

An equation analogous to (96) has been defined for $A_E(G)$ but instead of the total number of bonds B, it incorporates the total number of *connections*, N_C:

$$A_E(G) = 2N_C \tag{109}$$

The connections are the simplest-graph invariant which is composed of both vertices and edges. Specifically, they are graph fragments containing two adjacent edges. For molecular graphs without multiedges the connections equal the number of C_3 (propane) chain fragments which can be embedded into the carbon skeleton of a given molecule. The number of graph connections has also been used as a topological index in molecular branching studies by Gordon and Scantlebury (1964).

The first two information indices defined on the edge adjacency matrix of a graph (Bonchev and Mekenyan, 1982 a) are: *information indices on the equality and magnitude of edge adjacency matrix elements*, $^{E}\bar{I}^{E}_{adj}(G)$ and $^{E}\bar{I}^{M}_{adj}(G)$, respectively:

$$^{E}\bar{I}^{E}_{adj} = - \frac{2N_C}{B^2} \, \text{lb} \, \frac{2N_C}{B^2} - \frac{(B^2-2N_C)}{B^2} \, \text{lb} \, \frac{(B^2-2N_C)}{B^2} \tag{110}$$

$$^{E}\bar{I}^{M}_{adj} = \text{lb} \, A_E = \text{lb} \, 2N_C = 1 + \text{lb} \, N_C \tag{111}$$

For graphs G_1 and G_2 it is obtained by eq (110) and (111): ${}^E\bar{I}^E_{adj}$ (G_1) = 0.9755 bits, ${}^E\bar{I}^M_{adj}$ (G_1) = 4.3219 bits,

${}^E\bar{I}^E_{adj}$ (G_2) = 0.9925 bits, and ${}^E\bar{I}^M_{adj}$ (G_2) = 4.4594 bits.

The second pair of indices is defined analogously to eq (101) and (102). These are *information indices on the edge degree and magnitude* of a graph, ${}^E\bar{I}^E_{adj, \ deg}$ (G) and ${}^E\bar{I}^M_{adj,deg}$ (G), respectively:

$$
{}^E\bar{I}^E_{adj,deg} \ (G) = - \sum_i \frac{N_{e_i}}{B} \ \text{lb} \ \frac{N_{e_i}}{B} \tag{112},
$$

$$
{}^E\bar{I}^M_{adj,deg} \ (G) = - \sum_i \frac{e_i}{2N_C} \ \text{lb} \ \frac{e_i}{2N_C} \tag{113},
$$

where N_{e_i} is the number of edges of the same edge degree, e_i.

The numerical values of the above two indices for graphs G_1 and G_2 have been calculated from the corresponding partitions:

${}^E_P{}^E_{adj,deg}$ (G_1) = 7(4,1,1,1); ${}^E\bar{I}^E_{adj,deg}$ (G_1) = 1.6645 bits

${}^E\bar{I}^M_{adj,deg}$ (G_1) = 20(5,3,3,3,3,2,1);

${}^E\bar{I}^M_{adj,deg}$ (G_1) = 2.6905 bits

$$E_P{}^E_{adj,deg} (G_2) = 7(3,2,2);$$

$$E_{\bar{I}}{}^E_{adj,deg} (G_2) = 1.5567 \text{ bits}$$

$$E_P{}^M_{adj,deg} (G_2) = 22(4,4,4,3,3,2,2);$$

$$E_{\bar{I}}{}^M_{adj,deg} (G_2) = 2.7544 \text{ bits}$$

Besides the partitions of the adjacency matrix elements according to their equality or magnitude, the subset of units in $A(G)$ and $A_E(G)$ can be divided additionally into classes of orbital equivalency. The elements which belong to the same orbit of the graphs automorphism group are qualified as equivalent.

In case of $A(G)$ the orbits contain the graph edges ($a_{ij} = 1$ each stands for one of the graph edges). We have already presented in subsection 3.B.i such an index on the edge orbital equivalency, $E_{\bar{I}}{}_{ORB}$. Dealing with an analogous orbital partition of the $A_E(G)$ non-zero entries, however, a new *information index on the orbital equivalence of the graph connections*, $\bar{I}_{ORB,CONN}$, can be defined:

$$\bar{I}_{ORB,CONN} = - \sum_{i=1}^{k} \frac{N_{c,i}}{N_c} \text{ lb } \frac{N_{c,i}}{N_c} \tag{114}$$

where $N_{c,i}$ is the number of connections in the ith orbit, N_c is their total number, and k stands for the total number of orbits. Such an index has been introduced by Bertz (1981) for multigraphs as a component of a general index of molecular complexity. It can be determined for graphs without multiedges and for directed graphs, as well.

Examples (See $A_E(G_1)$ and $A_E(G_2)$)

Connection orbits of G_1: $\{E_1E_2,\ E_1E_3,\ E_2E_3\}$, $\{E_1E_4,\ E_2E_4,$ $E_3E_4\}$, $\{E_4E_5\}$, $\{E_4E_7\}$, $\{E_5E_6\}$, $\{E_5E_7\}$

Connection orbital partition:

$$P_{ORB,CONN}(G_1) = N_c\{N_{c,1}, N_{c,2},\ \ldots,\ N_{c,k}\} = 10\{3,3,1,1,1,1\}$$

$$\bar{I}_{ORB,CONN}(G_1) = -\frac{3}{5}\ lb\ \frac{3}{10} - \frac{2}{5}\ lb\ \frac{1}{10} = 2.3710\ bits$$

Connection orbits of G_2: $\{E_1E_2,\ E_1E_3\}$, $\{E_2E_3\}$, $\{E_2E_4,\ E_3E_4\}$, $\{E_2E_5,\ E_3E_6\}$, $\{E_4E_5,\ E_4E_6\}$, $\{E_5E_7,\ E_6E_7\}$

$$P_{ORB,CONN}(G_2) = 11\{2,2,2,2,2,1\}$$

$$\bar{I}_{ORB,CONN}(G_2) = -\frac{10}{11}\ lb\ \frac{2}{11} - \frac{1}{11}\ lb\ \frac{1}{11} = 2.5503\ bits$$

It should be mentioned here that Valentinuzzi and Valentinuzzi (1962) have presented examples for a similar extension of the orbital information concept to graph fragments that have been more complicated than vertices and edges. The three isomeric molecules of o-, m-, and p-aminobenzoic acid have been expressed by the following simplified graphs (FIG. 19)

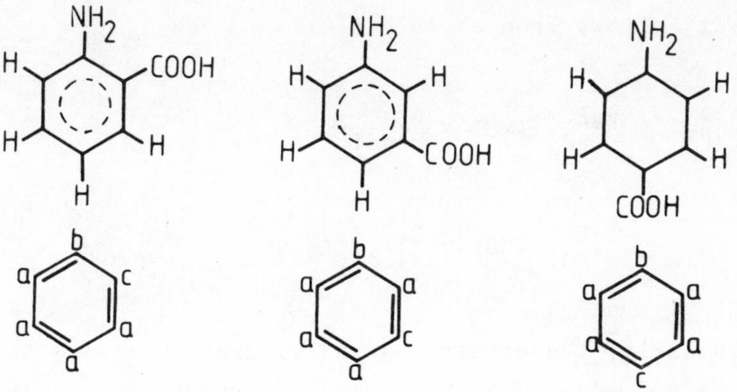

FIG.19. Simplifed graphs of the o-, m-, and p- amino-
benzoic acid molecules

The orbital equivalency of the connections in these
graphs, e.g.

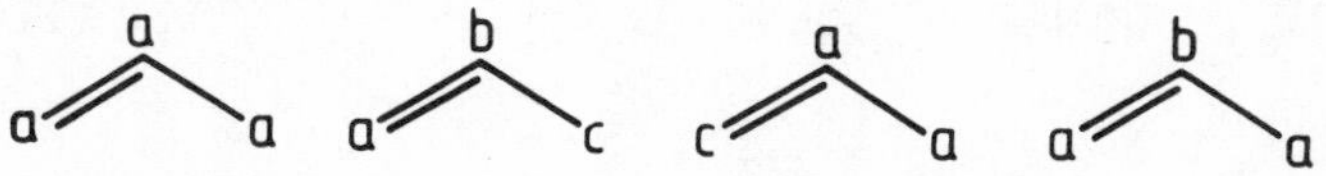

, etc.

have been treated by these authors.

v. Information Indices Based on the Vertex and Edge
Adjacency Matrices of Multigraphs

All of the indices defined on the basis of vertex and edge
adjacency matrices of graphs by means of eq (100-102) and
(110-114) can also be defined for multigraphs. In most
cases exact relations can be found (Bonchev and Mekenyan,
1982 a) between the multigraph *MG* and its parent graph *G*
in which a simple edge stands for each multiedge of MG.

$A(MG)$ and $A_E(MG)$ contain at least one entry a_{ij} = 2 or 3. Hence the subset of units in the parent graph G splits and it follows from eq (97, 110):

$$^{V}\bar{I}^{E}_{adj} \ (MG) > {}^{V}\bar{I}_{adj} \ (G) \tag{115}$$

$$^{E}\bar{I}^{E}_{adj} \ (MG) > {}^{E}\bar{I}_{adj} \ (G) \tag{116}$$

Similarly, the presence of multiedges increases the total vertex and edge adjacency A and A' (eq 96, 10%). Hence, also from eq (100) and (111), one arrives to the inequalities

$$^{V}\bar{I}^{M}_{adj} \ (MG) > {}^{V}\bar{I}_{adj} \ (G) \tag{117}$$

$$^{E}\bar{I}^{M}_{adj} \ (MG) > {}^{V}\bar{I}_{adj} \ (G) \tag{118}$$

No such inequalities can be found for the information indices on the vertex - or edge-degree equality, $^{V}\bar{I}^{E}_{adj,deg}$ and $^{E}\bar{I}^{E}_{adj,deg}$, respectively. Depending on the change in the vertex or edge partitions on the transformation of the parent graph G into a multigraph MG, these two indices can increase, decrease, or remain constant.

A multigraph information content greater than the corresponding information content of the parent graph is again obtained when dealing with the information on the edge total adjacency partition into edge degrees (or otherwise, information on the magnitudes of edge degrees, eq 113):

$$E_{\bar{I}}^M{}_{adj,deg} (MG) > E_{\bar{I}}^M{}_{adj,deg} (G) \tag{119}$$

Example:

edge degree partition:

$$E_P^M{}_{adj,deg} (MG_2) = A'\{e_1, e_1{}'e_2, e_3, e_4,$$
$$e_5, e_6, e_7, e_8\} =$$
$$= 28\{3,3,5,5,4,3,3,2\}$$

$$E_{\bar{I}}^M{}_{adj,deg} (MG_2) = 2.9417 > E_{\bar{I}}^M{}_{adj,deg} (G_2) = 2.7544 \text{ bits}$$

One may expect an inequality similar to (115) to hold for the information on the magnitudes of vertex degree, $v_{\bar{I}}^M{}_{adj,deg}$, eq (102). The analysis, however, shows that this is only a dominant trend with exceptions for small multigraphs with a single double edge, only.

Example:

$$v_P^M{}_{adj,deg} (MG) = 8\{2,3,2,1\}$$

$$v_P^M{}_{adj,deg} (G) = 6\{1,2,2,1\}$$

$$v_{\bar{I}}^M{}_{adj,deg} (MG) = 1.9056 < v_{\bar{I}}^M{}_{adj,deg} (G) = 1.9183 \text{ bits}$$

Finally, the orbital information on graph connections has also been proved to be less than that of the respective multigraphs:

$$\bar{I}_{ORB,CONN} (MG) > \bar{I}_{ORB,CONN} (G) \qquad (120)$$

Example:

connection orbits of MG_2:
$\{E_1 E_1', E_1' E_1\}$, $\{E_1 E_2, E_1 E_3, E_1' E_2,$
$E_1' E_3\}$, $\{E_2 E_3\}$, $\{E_2 E_4, E_3 E_4\}$,
$\{E_2 E_5, E_3 E_6\}, \{E_4 E_5, E_4 E_6\}, \{E_5 E_7,$
$E_6 E_7\}$

$$P_{ORB,CONN} (MG_2) = 15\{4,2,2,2,2,2,1\}$$

$$\bar{I}_{ORB,CONN} (MG_2) = 2.7069 > \bar{I}_{ORB,CONN} (G_2) = 2.5503 \text{ bits}$$

The above comparison between the information indices defined by the vertex and edge adjacency matrices of graphs and multigraphs indicates which one of the indices derived is a convenient means for the characterization of molecules having multiple bonds. Simultaneonsly, it is a necessary step in the selection of indices to construct a general index of molecular complexity (See Chapter V). Evidently, indices like $^{V}\bar{I}^{E}_{adj}$, $^{E}\bar{I}^{E}_{adj}$, $^{V}\bar{I}^{M}_{adj}$, $^{E}\bar{I}^{M}_{adj}$, $^{E}\bar{I}^{M}_{adj,deg}$, and $\bar{I}_{ORB,CONN}$ can be used for such purposes since, as shown by inequalities (115-130), they correctly qualify multigraphs as more complex structures than graphs.

Conversely, indices $^{V}\bar{I}^{E}_{adj,deg}$, $^{E}\bar{I}^{E}_{adj,deg}$, and $^{V}\bar{I}^{M}_{adj,deg}$ are not recommended for use in the estimation of general molecular complexity.

A similar selection can be made for the orbital and chromatic indices, introduced in 3.B, for graph vertices and edges. Proceeding from eq (77-93), and the analysis presented there, a conclusion is drawn that a single index of this type can be used to construct a general index of molecular complexity, namely, $^{E}\bar{I}_{CHR}$.

vi. Information Indices Based on the Vertex and Edge Adjacency Matrices of Directed Graphs

All the nine information indices introduced for graphs and multigraphs on the basis of their vertex and edge adjacency matrices can also be defined for directed graphs. The comparison between the corresponding indices for directed graphs and their parent undirected graph result in the following inequalities (Bonchev and Mekenyan, 1982 a):

$$^{V}\bar{I}^{M}_{adj} \text{ (DG)} < ^{V}\bar{I}^{M}_{adj} \text{ (G)} \tag{121}$$

$$^{E}\bar{I}^{M}_{adj} \text{ (DG)} < ^{E}\bar{I}^{M}_{adj} \text{ (G)} \tag{122}$$

$$I_{ORB,CONN} \text{ (DG)} \geqslant I_{ORB,CONN} \text{ (G)} \tag{123}$$

The other six indices: $^{V}\bar{I}^{E}_{adj}$, $^{E}\bar{I}^{E}_{adj}$, $^{V}\bar{I}^{E}_{adj,deg}$, $^{E}\bar{I}^{E}_{adj,deg}$, $^{V}\bar{I}^{M}_{adj,deg}$, $^{E}\bar{I}^{M}_{adj,deg}$, could increase, or decr-

118

ease,or (for some indices) remain constant when the non-oriented graph is transformed into an oriented one.

Examples:

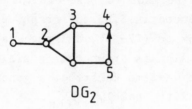 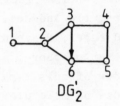

DG$_2$ DG$_2'$

Taking into account that the *A(DG)* entries (a_{ij}) equal 1 only for an arc incident *out* of i into j ($a_{ij} = 0$ when there is an arc incident *in* of i out of j) the following partitions result:

$$^V P_{adj}^E \ (DG_2) = 49(13,36)$$

$$^V P_{adj}^M \ (DG_2) = 13(13 \times 1)$$

$$^V \bar{I}_{adj}^E \ (DG_2) = 0.8346 < \ ^V \bar{I}_{adj}^E \ (G_2) = 0.9641 \text{ bits}$$

$$^V \bar{I}_{adj}^M \ (DG_2) = 3.7004 < \ ^V \bar{I}_{adj}^M \ (G_2) = 3.8074 \text{ bits}$$

$$^V P_{adj,deg}^M \ (DG_2) = 13(1,3,3,1,2,3)$$

$$^V P_{adj,deg}^M \ (DG_2') = 13(1,3,2,2,2,3)$$

$$^{v}\overline{I}^{M}_{adj,deg} (DG_2) = 2.4493 < {}^{v}\overline{I}^{M}_{adj,deg} (G_2) = 2.5027$$

$$^{v}\overline{I}^{M}_{adj,deg} (DG'_2) = 2.5074 > {}^{v}\overline{I}^{M}_{adj,deg} (G_2) = 2.5027$$

Examples for inequality (123) are both DG_2 and DG'_2 as compared with G_2 while equality (122) is illustrated below with the directed graph DG''_2.

$$G_2 \qquad\qquad DG''_2$$

The above considerations may be useful in the search for convenient indices, capable of comparing the general complexity of oriented and non-oriented graphs. Such a comparison is of importance when estimating the complexity of linear reaction mechanisms involving reversible, or/ and irreversible reaction steps (Bonchev and Temkin, 1982). A linear mechanism of a chemical reaction is regarded as more complex when many of its elementary steps are reversible. Hence, information indices are needed which should have a larger magnitude for nondirected graphs than for directed ones. As seen from ineq (121-123), only the indices on the magnitude of the vertex-and edge adjacency matrix elements, $^{v}\overline{I}^{M}_{adj}$ and $^{E}\overline{I}^{M}_{adj}$, respectively, meet this requirement. The information indices on the orbital equivalence of graph vertices, edges and connections are not appropriate to this aim.

D. INFORMATION INDEX ON THE INCIDENCE OF A GRAPH

The incidence matrix of a graph, $I(G)$, has been used to specify the total and mean information index on the graph incidence (Bonchev and Trinajstić, 1977), I_{inc} (G) and $\bar{I}_{inc}$ (G), respectively. The incidence matrix is a rectangular matrix representation of a graph G whose elements are

i_{kj} = 1 if the edge E_j is incident to the vertex V_k,

i_{kj} = 0 otherwise

Example:

$$I(G_1) = \begin{array}{c|ccccccc} & E_1 & E_2 & E_3 & E_4 & E_5 & E_6 & E_7 \\ \hline V_1 & 0 & 1 & 0 & 0 & 0 & 0 & 0 \\ V_2 & 1 & 1 & 1 & 1 & 0 & 0 & 0 \\ V_3 & 0 & 0 & 0 & 1 & 1 & 0 & 1 \\ V_4 & 0 & 0 & 0 & 0 & 1 & 1 & 0 \\ V_5 & 0 & 0 & 0 & 0 & 0 & 1 & 0 \\ V_6 & 1 & 0 & 0 & 0 & 0 & 0 & 0 \\ V_7 & 0 & 0 & 1 & 0 & 0 & 0 & 0 \\ V_8 & 0 & 0 & 0 & 0 & 0 & 0 & 1 \end{array}$$

The following equations have been derived (Bonchev and Mekenyan, 1982 a):

$$I_{inc} (G) = NB \ \text{lb} \ N - B(N-2) \ \text{lb} \ (N-2) - 2B \tag{124}$$

$$\bar{I}_{inc}\ (G)\ =\ 1b\ N\ -\ \frac{N-2}{N}\ 1b\ (N-2)\ -\ \frac{2}{N} \tag{125}$$

proceeding from the partition of the NxB set into two subsets, those of units (2B) and zeros (NB-2B), respectively.

For G_1 and G_2 one obtains:

$$N = 8,\ B = 7,\ I_{inc}\ (G_1)\ =\ 45.4316,\ \bar{I}_{inc}\ (G_1)\ =\ 0.8113$$
$$\text{bits}$$

$$N = 6,\ B = 7,\ I_{inc}\ (G_2)\ =\ 38.5684,\ \bar{I}_{inc}\ (G_2)\ =\ 0.9183$$
$$\text{bits}$$

Equation (125) is to be noted for the lack of dependence of the mean graph incidence information on the number of edges in the graph. This is due to the fact that each edge contributes two units to the subset of units which makes the probability for a randomly chosen element to belong to this subset constant for a graph and derivable multigraphs: $p_1 = 2B/BN = 2/N = $ constant. As a consequence, a multigraph and its parent graph have the same value of this information index. The total information on graph incidence, however, is always greater in multigraphs than in graphs:

$$\bar{I}_{inc}\ (MG)\ =\ \bar{I}_{inc}\ (G) \tag{126}$$

$$I_{inc}\ (MG)\ -\ I_{inc}\ (G)\ =\ N\ 1b\ N\ -\ (N-2)\ 1b\ (N-2)-2>0 \tag{127}$$

In the case of directed graphs the definition of the incidence matrix changes so as to distinguish between an arc E_j incident *out* of the vertex V_k *(i_{kj} = + 1)*, and an arc E_j incident *into* the vertex V_k (i_{kj} = - 1). Thus, the subset of non-zero entries in the incidence matrix of the parent nondirected graph splits into two such subsets of positive and negative non-zero entries, for the derived directed graphs. A third part of these non-zero entries, referring to nondirected edges, turns into zero entries a_{kj} = 0, because in these cases the edge E_j is simultaneously incident out of and into the vertex V_k. As a result of these changes the information on the incidence of a directed graph:

$$\bar{I}_{inc} \text{ (DG)} = -\frac{N_o}{NB} \text{ lb } \frac{N_o}{NB} - \frac{N_{+1}}{NB} \text{ lb } \frac{N_{+1}}{NB} - \frac{N_{-1}}{NB} \text{ lb } \frac{N_{-1}}{NB} \qquad (128)$$

can be greater, as well as less, than that of the parent undirected graph. Therefore, this information index cannot be used to compare nondirected and directed graphs.

E. INFORMATION INDICES BASED ON THE CYCLE MATRICES OF A GRAPH

The *(edge) cycle matrix* C of a graph G is defined by its elements

$$c_{ij} = + 1, \quad \text{if the edge } E_j \text{ belongs to the } j \text{ th}$$
$$\text{cycle in } G,$$
$$c_{ij} = 0, \quad \text{otherwise}$$

Example:

C				
E	C_1	C_2	C_3	$c_i(E)$
E_1	0	0	0	0
E_2	1	0	1	2
E_3	1	0	1	2
$C(G_2) = \quad E_4$	1	1	0	2
E_5	0	1	1	2
E_6	0	1	1	2
E_7	0	1	1	2

cycles: $C_1\{E_2,E_3,E_4\}$,

$C_2\{E_4,E_5,E_7,E_6\}$

$C_3\{E_2,E_5,E_7,E_6,E_3\}$

As seen the cycle matrix $C(G)$ is a rectangular edge/ cycle "incidence" matrix. The sum over all entries in the ith row specifies what is called here an *edge cyclic degree*, $c_i(E)$:

$$c_i(E) = \sum_{j=1}^{x} c_{ij} \qquad (129)$$

where x stands for the total number of cycles in G.

The sum over all edge cycle degrees is called here a *total edge cyclicity* of the graph, $C_E(G)$:

$$C_E(G) = \sum_{i=1}^{B} c_i(E) = \sum_{i=1}^{B} \sum_{j=1}^{x} c_{ij} \qquad (130)$$

As seen from eq (130), $C_E(G)$ increases rapidly with the total number of cycles and can be used as a general characteristic of molecular cyclicity.

Similarly to the edge adjacency matrix, four information indices can be defined on the (edge) cycle matrix of a graph G. These are (Bonchev and Mekenyan, 1982 a) the information on the equality or magnitude of the (edge)

cycle matrix elements, $^{E}\overline{I}^{E}_{cycl}$ (G) and $^{E}\overline{I}^{M}_{cycl}$ (G), as well as the information on the equality or magnitude of the edge cycle degrees, $^{E}\overline{I}^{E}_{cycl,deg}$ (G) and $^{E}\overline{I}^{M}_{cycl,deg}$ (G), respectively.

Instead of presenting details about these indices, we discuss below the alternative four information indices, based on the *vertex cycle matrix* of a graph $C_v(G)$. (Information on such a matrix is lacking in the graph-theoretical literature, but it can be introduced analogously to the edge cycle matrix). The reason for the consideration of this matrix is its importance for technological calculations, Specifically, the *vertex cycle degrees*, $c_i(V)$, are of interest since they indicate the most significant devices in a technological scheme. The cost of the global calculations of such schemes may rise tremendously when starting with a randomly selected vertex of the technological graph. On the other hand, the *total vertex cyclicity* of the graph, $C_v(G)$, can be used, together with some information indices defined below, as a topological index for measuring the complexity of such technological graphs (Bonchev and Mekenyan, 1982 b):

$$c_i(V) = \sum_{j=1}^{x} c_{ij} \tag{131}$$

$$C_v(G) = \sum_{i=1}^{N} \quad c_i(V) = \sum_{i=1}^{N} \sum_{j=1}^{x} c'_{ij} \tag{132}$$

Here c'_{ij} stands for the elements of the vertex cycle matrix.

Example:

V	C_1	C_2	C_3	$c_i(V)$
V_1	0	0	0	0
V_2	1	0	1	2
V_3	1	1	1	3
V_4	0	1	1	2
V_5	0	1	1	2
V_6	1	1	1	3

$$C_v(G_2) =$$

$$C_1 = \{V_2, V_3, V_6\},$$

$$C_2 = \{V_3, V_4, V_5, V_6\}$$

$$C_3 = \{V_2, V_3, V_4, V_5, V_6\}$$

$$C_v(G_2) = \sum_{i=1}^{6} c_i(V) = 12$$

The information indices on the equality and magnitude of the vertex cycle matrix elements, $^{v}\bar{I}^{E}_{cycl}(G)$ and $^{v}\bar{I}^{M}_{cycl}(G)$, respectively, are defined as follows:

$$^{v}\bar{I}^{E}_{cycl}(G) = -\frac{N_1}{NX} \text{lb} \frac{N_1}{NX} - \frac{N_0}{NX} \text{lb} \frac{N_0}{NX} \qquad (133),$$

$$^{v}\bar{I}^{M}_{cycl}(G) = \text{lb} \, N_1 = \text{lb} \, C_v(G) \qquad (134)$$

where N and X are the total number of vertices and cycles, respectively, and the numbers of units and zeros in $C_v(G)$ is denoted by N_1 and N_0, respectively.

The vertex partition according to their cycle degrees, $c_i(V)$ specifies a third index (an information index on the equality of the vertex cycle degrees):

$$^v\bar{I}^E_{cycl,deg} (G) = - \sum_{j=0}^{X} \frac{N_j}{N} \text{ lb } \frac{N_j}{N} \tag{135}$$

where N_j is the number of vertices having the same cycle degree $c_j(G)$.

The fourth index of this series, the information on the magnitude of vertex cycle degrees, is based on the partition of the total vertex cyclicity of the graph (eq 132) into individual vertex cycle degrees:

$$^v\bar{I}^M_{cycl,deg} (G) = - \sum_{i} \frac{c_i(V)}{C_V(G)} \text{ lb } \frac{c_i(V)}{C_V(G)} \tag{136}$$

Examples (See $C_v(G_2)$ above):

$$N = 6, \ X = 3, \ N_0 = 6, \ N_1 = 12$$

$$^v\bar{I}^E_{cycl} (G_2) = - \frac{12}{6.3} \text{ lb } \frac{12}{6.3} - \frac{6}{6.3} \text{ lb } \frac{6}{6.3} = 0.9183 \text{ bits}$$

$$^v\bar{I}^M_{cycl} (G_2) = \text{ lb } 12 = 3.5850 \text{ bits}$$

$$c_1(V) = 0, \ c_2(V) = c_4(V) = c_5(V) = 2, \ c_3(V) = c_6(V) = 3$$

$$^v\bar{I}^E_{cycl,deg}(G_2) = - \frac{1}{6} \text{ lb } \frac{1}{6} - \frac{3}{6} \text{ lb } \frac{3}{6} - \frac{2}{6} \text{ lb } \frac{2}{6} = 1.4591 \text{ bits}$$

$$^{v}I^{M}_{cyc1,deg} (G_2) = - 3x \frac{2}{12} lb \frac{2}{12} - 2x \frac{3}{12} lb \frac{3}{12} = 2.2925 \text{ bits}$$

The information indices on the equality and magnitude of the vertex cycle degrees can be recommended as convenient measures of molecular cyclicity (or of the complexity of cyclic molecules).

The comparison of the mean information indices on the vertex cycle matrix of graphs and derived multigraphs, as well as directed graphs, indicates the existence of an exact relation only for the index on the magnitude of the matrix elements:

$$^{v}\bar{I}^{M}_{cyc1} (MG) > {}^{v}\bar{I}^{M}_{cyc1} (G) \geqslant {}^{v}\bar{I}^{M}_{cyc1} (DG) \tag{137}$$

The other three indices, defined by eq (133,135,136), can increase or decrease or remain constant when transforming the parent graph into a multigraph, or a directed graph.

F. INFORMATION INDICES BASED ON THE DISTANCE MATRIX OF A GRAPH

The distance matrix of a graph G, $D(G)$ has already been discussed in Subsection 3.A.v in connection with the definition of the total distance of a graph, the so-called Wiener number (eq 71).

Information indices on the distances in the graph have been introduced by Bonchev and Trinajstić (1977), considering all the matrix elements of distance matrix d_{ij} as

elements of a finite probability scheme associated with
the graph in question. Let the distance of a value l
appear $2k_l$ times in the distance matrix, where $l \leqslant 1 \leqslant$
$\leqslant l_m$. Then N^2 matrix elements d_{ij} will be partitioned into
$m + 1$ groups, where m is the highest value of l, and
group $m + 1$ will contain N zeros which are the diagonal
matrix elements. A certain probability for a randomly
chosen distance d_{ij} to be in the lth group can be associated with each one of these $(m + 1)$ groups:

$$p_1 = 2k_1/N^2, \qquad p_o = N/N^2 = 1/N,$$

Distances:	0	1	2	3	...	m
Number of distances:	N	$2k_1$	$2k_2$	$2k_3$	...	$2k_{\overline{m}}$
Probability:	$\dfrac{1}{N}$	$\dfrac{2k_1}{N^2}$	$\dfrac{2k_2}{N^2}$	$\dfrac{2k_3}{N^2}$	...	$\dfrac{2k_m}{N^2}$.

The *mean and total information on distance equality* of
a given graph I_D^E (G), is then, according to Eq(6) and (7):

$$\bar{I}_D^E \ (G) = - \frac{1}{N} \ \text{lb} \ \frac{1}{N} - \sum_{1=1}^{m} \ \frac{2k_1}{N^2} \ \text{lb} \ \frac{2k_1}{N^2}, \tag{138}$$

$$I_D^E \ (G) = N^2 \text{lb} \ N^2 - N\text{lb} \ N - \sum_{1=1}^{m} \ 2k_1\text{lb} \ 2k_1 \tag{139}$$

Since D (G) is a symmetric matrix, one can consider, for
simplicity of discussion, only the upper triangular submatrix which does indeed preserve all the properties of the information measure. In that case, we have the following
expressions for the mean and total information on distan-

ce equality:

$$\bar{I}_D^E(G) = - \Sigma \frac{2k_1}{N(N-1)} \; lb \; \frac{2k_1}{N(N-1)}, \tag{140}$$

$$I_D^E(G) = \frac{N(N-1)}{2} \; lb \; \frac{N(N-1)}{2} - \Sigma \; k_1 lb \; k_1 \tag{141}$$

where $N(N-1)/2$ is the total number of upper off-diagonal elements in D.

The *information index on the magnitude of distances* in a given graph, I_D^M, has also been defined. It is in fact the information on the partitioning Wiener number into groups of distances of the same magnitude. Since the Wiener number is given by

$$W = \Sigma_1 \; lk_1 \tag{142}$$

and following Eq (6) and (7), we obtain

$$I_D^M = W \; lb \; W - \Sigma_1 \; k_1 l \; lb \; l \tag{143}$$

and

$$\bar{I}_D^M = - \Sigma_1 \; k_1 \frac{1}{W} \; lb \; \frac{1}{W}. \tag{144}$$

Examples (See $D(G_1)$ and $D(G_2)$ in subsection 3.A.v)

$$W(G_1) = 7.1 + 10.2 + 8.3 + 3.4 = 63$$

$$\bar{I}_D^E(G_1) = -\frac{1}{4} \text{ lb } \frac{1}{4} - \frac{5}{14} \text{ lb } \frac{5}{14} - \frac{2}{7} \text{ lb } \frac{2}{7} - \frac{3}{28} \text{ lb } \frac{3}{28} =$$

$$= 1.8922 \text{ bits}$$

$$\bar{I}_D^M(G_1) = -\frac{1}{9} \text{ lb } \frac{1}{63} - \frac{20}{63} \text{ lb } \frac{2}{63} - \frac{8}{21} \text{ lb } \frac{1}{21} - \frac{12}{63} \text{ lb } \frac{4}{63} =$$

$$= 4.6751 \text{ bits}$$

$$W(G_2) = 7.1 + 6.2 + 2.3 = 25$$

$$\bar{I}_D^E(G_2) = -\frac{7}{15} \text{ lb } \frac{7}{15} - \frac{2}{5} \text{ lb } \frac{2}{5} - \frac{2}{15} \text{ lb } \frac{2}{15} = 1.4295 \text{ bits}$$

$$\bar{I}_D^M(G_2) = -\frac{7}{25} \text{ lb } \frac{1}{25} - \frac{12}{25} \text{ lb } \frac{2}{25} - \frac{6}{25} \text{ lb } \frac{3}{25} = 3.7835 \text{ bits}$$

Simple equations have been obtained for the information index on the equality of graph of various classes of acyclic graphs. E.g. for chains (n-alkanes):

```
    o    o    o    o....o    o    o
    1    2    3        N-2  N-1   N
```

$$I_D^E(\text{chain}) = \frac{N(N-1)}{2} \text{ lb } \frac{N(N-1)}{2} - \sum_{k=1}^{N-1} k \text{ lb } k \qquad (145)$$

The number of all possible distributions of d_{ij}, i.e. the number of different I_D^E and I_D^M, increases rapidly with increasing number of vertices in the graph. This makes the two information indices appropriate for the discrimination of isomeric molecules. They can distinguish graphs having the same Wiener index but different frequency number of their distances. Thus, I_D^E and I_D^M have different values for all acyclic hydrocarbons with four to eight carbon atoms.

A second pair of *information indices* has been defined (Bonchev and Mekenyan, 1982 a) *on the equality and magnitude of the distance degrees* of a graph, $\bar{I}^E_{D,deg}(G)$ and $\bar{I}^M_{D,deg}(G)$, respectively. The distance degree of a vertex i, d_i, was considered in Subsection 3.A.vi in connection with the Balaban distance connectivity index. The graph vertices can be partitioned into subsets of different magnitude of d_i, thus forming the vertex distance degree equality partition, ${}^v P^E_{D,deg}(G) = N\{N_{d_1}, N_{d_2}, \ldots, N_{d_n}\}$, N_{d_i} being the cardinality of the ith subset of vertices having a distance d_i. On the other hand, the doubled Wiener index (or otherwise, the Rouvray index $R = 2W$) can be regarded partitioned into individual distance degrees: ${}^v P^M_{D,deg}(G) = 2W\{d_1, d_2, \ldots, d_N\}$, the latter being the distance degree magnitude partition. These partitions provide the formulation of the two indices mentioned above making use of the basic equation (6):

$$
{}^v \bar{I}^E_{D,deg}(G) = - \sum_{i=1}^{n} \frac{N_{d_i}}{N} \, lb \, \frac{N_{d_j}}{N} \tag{146}
$$

$$
{}^v \bar{I}^M_{D,deg}(G) = - \sum_{i=1}^{N} \frac{d_i}{2W} \, lb \, \frac{d_i}{2W} \tag{147}
$$

Examples (See the distance matrices $D(G_1)$ and $D(G_2)$ in Subsection 3.A.v.):

$$
{}^v \bar{P}^E_{D,deg}(G_1) = 8(4,2,1,1), \quad {}^v \bar{I}^E_{D,deg}(G_1) = 1.7500 \text{ bits}
$$

$$
{}^v P^M_{D,deg}(G_1) = 126(21,4\times17,15,2\times11),
$$

$$^V P^E_{D,deg}(G_2) = 6(3,2,1), \quad ^V \bar{I}^E_{D,deg}(G_2) = 1.4591 \text{ bits}$$

$$^V \bar{P}^M_{D,deg}(G_2) = 50(11,2 \times 9,3 \times 7),$$

$$^V \bar{I}^M_{D,deg}(G_2) = 2.5625 \text{ bits}$$

The information indices specified above on the distance matrix of a graph $D(G)$ are sensitive measures of molecular branching (Bonchev and Trinajstić, 1977; 1978; Bonchev et al., 1979 a) and cyclicity (Bonchev and Mekenyan, 1982 a). With a few exceptions $(^V \bar{I}^E_{D,deg})$ they change regularly with increasing branchning and cyclicity as seen from the following examples (FIG.20):

a	b	c

$$^V \bar{I}^E_D = 1.8464 \qquad\qquad 1.5219 \qquad\qquad 0.9710$$

$$^V \bar{I}^M_D = 3.1464 \qquad\qquad 3.1972 \qquad\qquad 3.2500$$

$$^V \bar{I}^M_{D,deg} = 2.2906 \qquad\qquad 2.2910 \qquad\qquad 2.2936$$

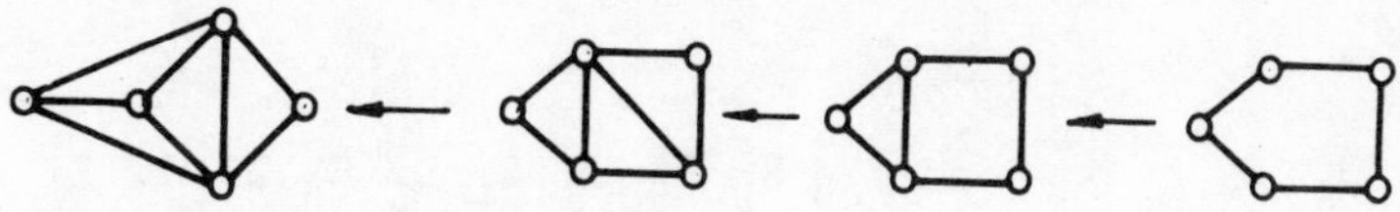

$$^v\bar{I}^E_D = 1. \qquad 0.9710 \qquad 0.8813 \qquad 0.7219$$

$$^v\bar{I}^M_D = 3.2402 \qquad 3.2359 \qquad 3.2389 \qquad 3.2516$$

$$^v\bar{I}^M_{D,deg} = 2.3219 \qquad 2.3163 \qquad 2.3066 \qquad 2.3046$$

FIG.20. Three information indices on the graph distances reflecting molecular branching and cyclicity

The above examples also indicate that, though measured by means of the same information index, the branching and cyclicity of molecules should be treated by two separate scales which overlap in case of the most branched and least cyclic molecules. The information index on the distance degree magnitudes, $^v\bar{I}^M_{D,deg}$, dispays an interesting behaviour; it increases with branching and decreases with cyclicity. This makes it important for some correlations with the π-electron energy of molecules which decreases with branching and mostly increases with cyclicity. The applications of the information indices on the graph distances to correlations with various molecular properties are discussed in the next chapter.

An analogous series of four indices can be specified (Bonchev and Mekenyan, 1982 a) on the basis of the *edge distance matrix*, $D_E(G)$, the sum of all its entries (called here *total edge distance* of the graph) D_E, or the sums of the entries in a row or column of this matrix (called here *edge distance degrees)*, $d_i(E)$.

Example

	E_1	E_2	E_3	E_4	E_5	E_6	E_7	$d_i(E)$
E_1	0	1	1	2	2	2	3	11
E_2	1	0	1	1	1	2	2	8
E_3	1	1	0	1	2	1	2	8
E_4	2	1	1	0	1	1	2	8
E_5	2	1	2	1	0	2	1	9
E_6	2	2	1	1	2	0	1	9
E_7	3	2	2	2	1	1	0	11

$$D_E(G_2) =$$

$$D_E(G_2) = 64$$

$${}^E\bar{I}_D^E(G_2) = 1.6388 \text{ bits}$$

$${}^E\bar{I}_D^M(G_2) = 5.2889 \text{ bits}$$

$${}^E\bar{I}_{D,deg}^E(G_2) = 1.5567 \text{ bits}$$

$${}^E\bar{I}_{D,deg}^M(G_2) = 2.7943 \text{ bits}$$

The *vertex* distance matrices of a graph and derived multigraphs coincide since the (vertex) distance is the shortest path connecting two vertices no matter whether this shortest path is single or multiple. Thus, in graphs below

the distance d_{14} is always three. Therefore, the four in-
dices $^{V}\bar{I}_{D}^{E}$, $^{V}\bar{I}_{D}^{M}$, $^{V}\bar{I}_{D,deg}^{E}$, and $^{V}\bar{I}_{D,deg}^{M}$, defined above on the
basis of this matrix, cannot distinguish between graphs
and multigraphs and have the same values for them.

The edge distance matrix of a multigraph, however, is
supplemented by a row and a column for each multiedge as
compared with the matrix of the parent graph. Therefore,
the information indices on the equality and magnitude of
the edge distance matrix entries, as well as those on the
equality and magnitude of the edge distance degrees will,
in general, have different values for a graph and the derived multigraphs. Inequalities, showing the increase in
the information indices on the magnitude of matrix elements and edge distances in multigraphs, as compared with
the parent graphs, have been proved (Bonchev and Mekenyan
1982 a):

$$^{E}\bar{I}_{D}^{M} (MG) > {}^{E}\bar{I}_{D}^{M} (G) \tag{148}$$

$$^{E}\bar{I}_{D,deg}^{M} (MG) > {}^{E}\bar{I}_{D,deg}^{M} (G) \tag{149}$$

Example:

$$^{E}P_{D}^{M} (MG_2) = 90(26 \times 1, \ 26 \times 2, \ 4 \times 3)$$

136

$$E_{P_{D,deg}^M} (MG_2) = 90(14, 2 \times 13, 2 \times 11, 10, 2 \times 9)$$

$$E_{\bar{I}_D^M} (MG_2) = 5.7027 > E_{\bar{I}_D^M} (G_2) = 5.2889 \text{ bits}$$

$$E_{\bar{I}_{D,deg}^M} (MG_2) = 2.9819 > E_{\bar{I}_{D,deg}^M} (G_2) = 2.7943 \text{ bits}$$

The eight information indices defined by the vertex and edge distance matrices cannot, however, be of use for directed graphs since in the general case infinite distances appear in these matrices. Pairs of vertices or edges, whose distance is infinite, are said to be unreachable.

Examples:

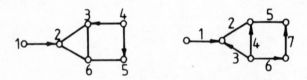

Vertices 1 and 4 are unreachable for the other four vertices since they are not end points for any arc. Arc 1 cannot reach arc 6 since it cannot reach its incident vertices 5 and 6.

G. AN INFORMATION INDEX BASED ON THE HOSOYA GRAPH DECOMPOSITION

The Hosoya non-adjacent numbers $P(G, k)$, expressing the number of ways k edges in graph G can be chosen so that no two of them should be adjacent, were introduced in Subsection 3.A.iv. Regarding $P(G,k)$ as elements of their sum partition, the Hosoya number of a graph G, $Z(G)$, a finite probability scheme, can be constructed. Thereby the information index on the Hosoya graph decomposition, $\bar{I}_Z(G)$, is defined (Bonchev et al., 1981 b):

$$\bar{I}_Z(G) = - \sum_{k=0}^{N/2} \frac{p(G;k)}{Z} \, lb \, \frac{p(G;k)}{Z} \tag{150}$$

Since for an acyclic graph Z is alternatively defined as the sum of the absolute values of coefficient in its characteristic polynomial, $\bar{I}_Z$ coincides in this case with the previously defined information for polynomial coefficients of the graph, $\bar{I}_{PC}$ (Bonchev and Trinajstić, 1977). *Examples* (See also examples in 3.A.iv)

$$P_Z(G_1) = 22\{1,7,11,3\}, \quad \bar{I}_Z(G_1) = 1.6203 \text{ bits}$$

$$P_Z(G_2) = 20\{1,7,10,2\}, \quad \bar{I}_Z(G_2) = 1.5784 \text{ bits}$$

$\bar{I}_Z$ index can also be calculated for multigraphs, which always have a higher such index, as compared with the parent graph (Bonchev and Mekenyan 1982 a):

$$\bar{I}_Z(MG) \geqslant \bar{I}_Z(G) \tag{151}$$

Example:

MG₂

$$P_Z(MG_2) = 27\{1,8,14,4\}$$

$$\bar{I}_Z(MG_2) = 1.6715 >$$

$$> \bar{I}_Z (G_2) = 1.5784 \text{ bits}$$

For directed graphs such an index (as well as its basis, the Hosoya number) could be defined solely under some special condition for the mutual directions of the pair of arcs but this possibility will not be explored here.

H. CENTRIC INFORMATION INDICES

i. Vertex Centric Indices

The lopping partition of the vertices in acyclic graphs, introduced by Balaban (1979) in order to define a centric topological index (See Subsection 3.A.vii), can also be used as a basis for a finite probability scheme. A *lopping centric information index*, $^V I_{C,L}(G)$, results in this way:

$$^V \bar{I}_{C,L}(G) = - \sum_{i=1}^{k} \frac{\delta_i}{N} \text{ lb } \frac{\delta_i}{N} \tag{152}$$

where δ_i is the number of vertices deleted in the ith step of the lopping procedure.

Example:

$$^V P_{C,L}(G_1) = 8(5,2,1)$$

$$^V \bar{I}_{C,L}(G_1) = 1.2988 \text{ bits}$$

A *radial centric information index* for both acyclic and cyclic graphs has been introduced by Bonchev et al. (1980c) on the basis of the classical definition for a graph centre (See Subsection 3.A.vii).

The radial vertex partition is an ordered set of the cardinalities N_{rj} of the subsets of vertices having the same radius r_j, beginning with the largest radius and ending with the central vertices:

$$^v P_{C,R} = N\{N_{jmax}, N_{jmax-1}, \ldots, N_{jmin}\}$$

$$^v \bar{I}_{C,R}(G) = \sum_j \frac{N_{r_j}}{N} \, 1b \, \frac{N_{r_j}}{N} \tag{153}$$

Examples:

r = 4: 1,5,6,7; r = 3: 2,4,8;
r = 2: 3

$$^v P_{C,R}(G_1) = 8(4,3,1)$$

$$^v \bar{I}_{C,R}(G_1) = 1.4056 \text{ bits}$$

r = 3: 1,4,5; r = 2; 2,3,6

$$^v P_{C,R}(G_2) = 6(3,3)$$

$$^v \bar{I}_{C,R}(G_2) = 1 \text{ bit}$$

More detailed centric partitions can be formed within the generalized graph centre concept (Bonchev et al., 1980 b, 1981 a). The need of such generalization, arose from the unrealistically high number of vertices qualified as central by the known definition in the case of cyclic graphs. This high number reduces sharply the efficiency of the notion of graph centre. Thus, all vertices of the graph shown below should be regarded as central since they have the same radius, r = 2:

The generalized concept is based on the distance matrix $D(G)$ and uses four criteria in a specified order. The central point(s) provide(s): (1) the smallest distance r_j to any other vertex: (2) the smallest sum of distances to all other vertices: (3) the smallest number of times the maximum distance occurs in the distance code, and (4) constancy in repeating criteria (1) - (3) to the pseudoc-entre graph containing only the vertices selected by the previous criteria and their incident edges.

As seen, the first criterion coincides with the known graph centre definition. The second criterion deals ac-tually with vertex distance degrees which have already being defined as the sum of the d_{ij}-entries in the ith row of $D(G)$ (See 3.A.vi).

The graph G_1 is shown below as an example where the subsequent application of criteria 1 and 2 is sufficient to determine the graph centre and to classify the graph vertices into a sequence of the classes of orbital equivalence reducing the number of central points to the minimum.

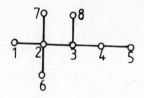

G_1

Vertices 1,5,6,7 have been found to have radius $r_i = 4$. Vertex 5, however, has a distance degree $d_5 = 21$ while for vertices 1,6, and 7 have $d_1 = d_6 = d_7 = 17$. Vertices 2,4, and 6 have the same radius, $r_i = 3$. Their distance degrees are, however, all different: $d_2 = 11$, $d_4 = 15$, $d_8 = 17$. Hence, the centric distance degree partition is

$$^v P_{C,DEG}(G_1) = 8(1,3,1,1,1,1) = {}^v P_{ORB}(G_1)$$

$$^v \bar{I}_{C,DEG}(G_1) = 2.4056 \text{ bits} = {}^v \bar{I}_{ORB}(G_1)$$

When criteria 1 and 2 fail to discriminate among several vertices,one can introduce a third criterion by comparing the distance codes of these vertices. The term *vertex distance code* is used for the shorthand notation of all entries in a row or column corresponding to a certain vertex: the entries are arranged in increasing order and, to save space, when a distance appears a number of times this number is written as a superscript. An example is shown below.

The vertex distance code is similar to the Randić atomic code (Randić 1979; Randić et al., 1979), written in the same concise manner but based on all possible paths (and not, as in the present case, on distances which are the shortest paths only) between the graph vertices. It is easily seen that the two codes are identical for acyclic

Distance matrix	r_i	d_i	distance code
0 1 2 3 2 1 2 2	3	13	$1^2 2^4 3^1$
1 0 1 2 1 2 3 3	3	13	$1^3 2^2 3^2$
2 1 0 1 2 3 4 4	4	17	$1^2 2^2 3^1 4^2$
3 2 1 0 1 4 5 5	5	21	$1^2 2^1 3^1 4^1 5^2$
2 1 2 1 0 3 4 4	4	17	$1^2 2^2 3^1 4^2$
1 2 3 4 3 0 1 1	4	15	$1^3 2^1 3^2 4^1$
2 3 4 5 4 1 0 2	5	21	$1^1 2^2 3^1 4^2 5^1$
2 3 4 5 4 1 2 0	5	21	$1^1 2^2 3^1 4^2 5^1$

graphs (trees).

In the above example, the first two criteria indicate
both vertices 1 and 2 as graph centres, but vertex 1 has
a different distance code from vertex 2. The largest radius, r_i = 3, occurs twice for vertex 2, but only once for
vertex 1; therefore, vertex 1 will be the centre accord ing
to the third criterion.

The centric distance code partition will then be:

$$^V P_{C,CODE}(G_{10}) = 8(1,2,2,1,1,1)$$

$$^V \bar{I}_{C,CODE}(G_{10}) = 2.5000 \text{ bits}$$

By applying in the given order criteria 1,2, and 3, we
obtain a number of vertices which constitute what is termed the *pseudocentre* of the graph. To obtain the final
reduction in the number of vertices in the centre, the
fourth criterion is applied. It consists in deleting from
the initial given graph all vertices but the pseudocentre
vertices, and all edges but those whose both endpoints are
pseudocentre vertices, resulting in a *pseudocentre graph;*
criteria 1-3 are applied in the same order to the

pseudocentre graph. The operation is repeated for the pseudocentre graph, and so on, until the pseudocentre remains unchanged after two consecutive applications of criteria 1-4. The result is called the *polycentre* of the graph.

An example for application of criterion 4 follows. In the example, the pseudograph contains four of the five initial vertices of the graph; application of criterion 4 eliminates from the pseudograph vertices 3 and 5, leaving the polycentre which contains vertices 1 and 2.

	1 2 3 4 5	r_i	d_r	Dist.code	Pseudograph r_i
	0 1 1 2 1	2	5	$1^3 2^1$	1
$D(G_{11}) =$	1 0 1 2 1	2	5	$1^3 2^1$	1
	1 1 0 1 2	2	5	$1^3 2^1$	2
	2 2 1 0 1	2	6	$1^2 2^2$	-
	1 1 2 1 0	2	5	$1^3 2^1$	2

G_{11}

The resulting partition is said to be a *complete centric partition*, and the respective index is called a *complete centric index*, $^v\bar{I}_{C,C}(G)$:

$$^v P_{C,C}(G_{11}) = 5(1,2,2)$$

$$^v \bar{I}_{C,C}(G_{11}) = 1.5219 \text{ bits}$$

The subsequent application of criteria 1 to 4 increases the discrimination of the graph vertices. It may be seen that the vertices grouped together become more and more

equivalent. The four criteria suffice in the great major-
ity of cases to provide a vertex *orbital* partitioning. The
last criterion used, which is sufficient to this aim,
therefore, results in a vertex centric index coinciding
with the vertex orbital index. The two kinds of indices
could, however, be discriminated since the centric procedure for finding the graph orbits orders them centroidally, starting with the outermost orbit and ending by the
innermost one, containing the central vertices.

This order can be taken into account by using the
Muirhead reordering procedure (See Hardy et al., 1934).
Partial sums are formed from each partition, by replacing
the second number in the partition by the sum of the first
two numbers, then replacing the third number in the sequence by the sum of the first three numbers, and so on.
Thus the differently ordered sequences of the same numbers
which cannot be discriminated by the information logarithmic function used in calculating topological indices, become
sequences of different numbers, in increasing order which
lead to distinct information indices. These modified partitions are called *Muirhead partitions*. To illustrate the
idea of Muirhead partitions one has first to make sure
that the number of terms in the partition equals their sum,
and this is done by adding as many zeroes as necessary
(for comparison's sake between all graphs having a certain
number of vertices, every structure is referred to the
graph in which all N vertices are distinct; hence, the
number of terms in the vertex partition should also be N,
the missing terms being added as zeroes). E.g. for G_1
having N = 8 the distance degree partition changes to the
Muirhead one by adding two zeroes:

$$^vP_{C,DEG}(G_1) = 8(1,3,1,1,1,1)$$

$$^{V}P_{C,DEG,M}(G_1) = 47(1,4,5,6,7,8,8,8)$$

$$^{V}\bar{I}_{C,DEG,M}(G_1) = 2.8573 \text{ bits}$$

There is no further coincidence with the vertex orbital information index for this particular example because this index is based on an unordered vertex partition.

An alternative series of four centric information indices can be defined replacing the radial vertex partition by the so - called *generalized radial vertex partition*. In the latter the graph vertices are ordered in subsets according to their average distance to the graph centre (polycentre) as defined after criterion 4. Criteria 2,3, and 4 can additionally be applied providing in general an increasing discrimination of the graph vertices within the resulting partitions: *generalized distance degree partition*, *generalized distance code partition* and *generalized complete centric partition*. *Generalized centric information indices* correspond to each of these four vertex partitions.

E.g. for graph G_{10} (See above) the central point is vertex 1. Vertices 2 and 6 are equally distant from it (d=1),

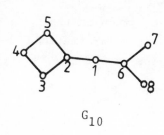

the same is true for vertices 3,5, 7 and 8 (d=2), and the most distant vertex 4 (d=3) is in an individual subset. The generalized radial vertex partition is then:

G_{10}

$$^{V}P_{C,GR}(G_{10}) = 8(1,4,2,1) \text{ differing}$$

from the centric radial vertex

partition $^{V}P_{C,R}(G_{10}) = 8(3,3,2)$, where the eight vertices are grouped according to their radius $r_i = 5(V_4, V_7, V_8)$, $4(V_3, V_5, V_6)$, and $3(V_1$ and $V_2)$, respectively. The application of criterion 2 leads to the generalized distance degree partition, $^V P_{C,G\ DEG}(G_{10}) = 8(1,2,2,1,1,1)$, in which the pair of vertices 3,5 and 7,8 are discriminated by their different distance degrees (17 and 21, respectively), and the same occurs with vertices 2 and 6 ($d_2 = 13$, $d_6 = 15$). The centric distance degree partition is again different from the generalized one: $^V P_{C,DEG}(G_{10}) = 8(3,2,1,2)$. Clearly the generalized centric information indices, calculated from the generalized centric vertex partitions, differ in many cases from the (classical) centric information indices.

ii. Edge Centric Indices

The hierarchical procedure of four criteria for the determination of the graph centre can be used in a similar manner in the search for the *central edge (s)* of the graph. (Bonchev and Gruncharov, 1982). The edge distance matrix $D_E(G)$ is of use in this case as well as the derived edge radius, $r_i(E)$, edge distance degree e_i, and edge distance code (See 3.F). Therefore, all vertex centric partitions and indices introduced in H.i could be similarly specified for edge centric partitions and indices.

Proceeding from the three partitions, given in the example below, the respective edge centric information indices are calculated: edge radial centric index:

$$^E \bar{I}_{C,R}(G_2) = 0.8631 \text{ bits}$$

Example:

G_2

E	1	2	3	4	5	6	7	$r_i(E)$	$d_i(E)$	distance code
1	0	1	1	2	2	2	3	3	11	$1^2 2^3 3^1$
2	1	0	1	1	1	2	2	2	8	$1^4 2^2$
3	1	1	0	1	2	1	2	2	8	$1^4 2^2$
$D_E(G_2)=4$	2	1	1	0	1	1	2	2	8	$1^4 2^2$
5	2	1	2	1	0	2	1	2	9	$1^3 2^3$
6	2	2	1	1	2	0	1	2	9	$1^3 2^3$
7	3	2	2	2	1	1	0	3	11	$1^2 2^3 3^1$

Partitions $(2,5);(2,2,3);(2,2,3)$

edge centric distance degree index:

$$^E\bar{I}_{C,DEG}(G_2) = 1.5567 \text{ bits}$$

edge centric distance code index:

$$^E\bar{I}_{C,CODE}(G_2) = {}^E\bar{I}_{C,DEG}(G_2)$$

Generalized edge centric partitions and indices can also be defined. In case of G_2 they coincide with the above partitions and indices, respectively.

The extended graph centre concept based on the four hierarchical criteria does not always suffice to reduce the number of central vertices and edges to the minimum. In case of some polycyclic graphs, several vertices (or edges) may be qualified as central though not all of them are topologically equivalent. Graph G_2 is such an example since three nonequivalent vertices (2,3, and 6) and edges (2,3, and 4) were found above to be central.

148

Two alternative approaches may be applied to terminate
the centric vertex and edge partitioning with a vertex and
edge orbital partition. The first one (Bonchev et al.,
1981 a) makes use of four additional criteria 1'to 4'
analogous to criteria 1 to 4 but based on paths instead of
distances. Vertex 2 is selected by this appoach as central
having paths of a maximum length of four (e.g. $V_2V_3V_4V_5V_6$)
while for vertices 3 and 6 this *path radius* is of length
five ($V_3V_4V_5V_6V_2V_1$). Similarly, edge 4 is found to be
central when applying criterion 1' since its path radius
is two (e.g. E_3E_1 or E_5E_7) while edges 2 and 3 have a path
radius of length three (e.g. for E_2: $E_5E_7E_6$).

The second approach (Bonchev and Cruncharov, 1982) makes
use of criteria 1 to 3 only applied to both vertices and
edges. Then an iterative procedure re-estimates the equivalence of the graph vertices by means of that of their
incident edges and vice versa up to their orbital equivalence. For graph G_2 vertices 3 and 6 are thus qualified
as central since they are incident to the central edge 4:

FIG.21. The black points denote the central vertices of
graph G_2 while the thick line indicates the central
edge in it.

The refined complete centric partitions obtained in
this way will coincide with the orbital vertex and edge
partitions. Being centroidally *ordered*, however, these
centric partitions may be reordered after Muirhead resulting thus in different information indices. Thus, the
complete edge centric partition of G_2 changes from

(2,2,3) to (1,1,2,2,1), which is actually the edge orbital partition, and then to the Muirhead edge centric partition 34(1,2,4,6,7,7,7). The information centric index calculated on this basis is $^E\bar{I}_{C,CM}(G_2)$ = 2.6032 bits.

iii. Centric Information Indices for Multigraphs

The centric information indices are based on the distance matrix of the graph. Due to this, all the conclusions made in Subsection 3.F. for the information indices on the graph distances are also valid for the centric indices. The latter cannot always be specified for directed graphs since some vertices in them could be unreachable which introduces infinite distances in the distance matrix. Similarly, a multigraph and its parent graph have the same vertex distance matrix and the vertex centric information indices do not distinguish them. The edge distance matrix of a multigraph has additional rows and columns, as compared with that of the parent graph and, therefore, provides different values of the edge centric information indices. The mean information indices of this type cannot reflect the increased complexity of multigraphs having larger, the same, as well as lower values, than those of the parent graphs.

Example:

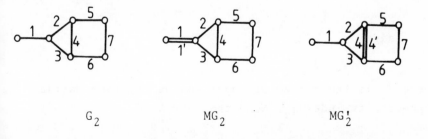

G_2 MG_2 MG'_2

radial

partition: 7(2,5)   8(3,5)   8(2,6)

$^E\bar{I}_{C,R}$:   0.8631   0.9544   0.8113

I. INFORMATION INDICES BASED ON COMBINED VERTEX-EDGE CONSIDERATIONS

i. Information Connectivity Indices

The molecular connectivity index χ_R of Randić (1975) has already been discussed in Subsection 3.A.iii. It was represented there as a sum over the partial connectivity indices χ_i of all i edges in the molecular graph. We note that the partial connectivity of an edge is a simple function of the degrees of its incident vertices.

A probability scheme can be constructed for the edges of the graph by partitioning them into k subsets depending on the equality of their partial connectivity indices χ_i. Making use of eq (6) an information index on the edges distribution in the graph according to their equivalence, $\bar{I}^E_{edge}$, as well as to their magnitude, $\bar{I}^M_{edge}$, has been defined (Bonchev et al., 1981 b) as

$$\bar{I}^E_{edge} = - \sum_{i=1}^{k} \frac{B_i}{B} \, lb \, \frac{B_i}{B} \, \text{bits,} \tag{154}$$

$$\bar{I}^M_{edge} = - \sum_{i=1}^{k} \frac{\chi_i}{\chi} \, lb \, \frac{\chi_i}{\chi} \, \text{bits,} \tag{155}$$

where B_i is the number of edges having the same partial connectivity index χ_i. Note that $B = \sum_{i=1}^{k} B_i$.

Example:

$$x_{12} = x_{26} = x_{27} = 0.5000$$
$$x_{38} = 0.5774, \; x_{4-5} = 0.7071$$
$$x_{23} = 0.2887, \; x_{3-4} = 0.4083$$

$$P^E_{edge}(G_1) = 7\{3,1,1,1,1\}; \; \bar{I}^E_{edge}(G_1) = 2.1281 \text{ bits}$$

$$P^M_{edge}(G_1) = 3.4815 \; \{3\times0.5, 0.5774, 0.7071, 0.2887, 0.4083\}$$

$$\bar{I}^M_{edge} = 2.7637 \text{ bits}$$

ii. Information Indices Based on First Neighbour Degrees
    and Edge Multiplicity

Such an index has been proposed by Sarkar, Roy et al.
(1978) for multigraphs and called "information content of
a multigraph". An unordered sequence called a *coordinate:*
$e_{x_1}, v_{x_1}; \; e_{x_2}, v_{x_2}; \; \ldots; \; e_{x_k}, v_{x_k},$ is assigned to vertex x_i
$(i = 1, 2, \ldots, N)$ where v_{x_r} is the degree of the vertex x_r
connected with x_i by means of e_{x_r} bonds. The multigraph
vertices are distributed into k equivalence classes, such
that vertices with same co-ordinate belong to the same
class. Hence, a finite probability scheme is constructed
and the information content of the multigraph [1]IC(MG) is
defined according to the Shannon equation (6).

152

Basak, Roy et al. (1979) modified the Sarkar - Roy index in the normalized form so as to eliminate the influence of the graph size. The new index has been termed *structural information content* (^{1}SIC):

$$^1SIC(MG) = \frac{\sum\limits_{i=1}^{k} p_i \, lb \, p_i}{lb \ N} \tag{156}$$

where N is the number of graph vertices.

^1SIC appears to reflect the Brillouin measure of information redundancy. ^1IC has also been modified by Raychaudhury et al. (1981) so as to measure the deviation of ^1IC from $^1IC_{max}$ and has been termed *complementary information content* (^1CIC). Another modification of ^1IC has been introduced by Ray et al. (1981 b) by replacing the number of vertices N in eq (156) by the number of bonds B. The resulting index is called *bonding information content*, ^1BIC. Using second and higher order neighbouring vertices and edge multiplicity, Roy et al. (1982) specified higher order information indices (^hIC, ^hSIC, ^hCIC, ^hBIC) together with a computer algorithm for their calculation.

Examples:

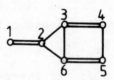

First order
Vertex coordinates
V_1:2,4
V_2:2,2;1,4;1,4
V_3:1,4;2,3;1,4

Second order
Vertex coordinates
V_1:(2,4;1,4); (2,4;1,4)
V_2:(1,4;2,3); (1,4;2,3)
V_3:(1,4;2,2); (2,3;1,3); (1,4;2,3)

$V_4 : 2,4;1,3$

$V_5 : 1,3;2,4$

$V_6 : 1,4;1,4;2,3$

$V_4 : (2,4;1,4); \ (2,4;1,4);$
$(1,3;2,4)$

$V_5 : (1,3;2,4); \ (2,4;1,4);$
$(2,4;1,4)$

$V_6 : (1,4;2,2); \ (1,4;2,3);$
$(2,3;1,3)$

Vertex partition first and second order coordinates

$$6\{1,1,2,2\}$$

$$^1IC(MG_2) = \ ^2IC(MG_2) = 1.9183 \text{ bits}$$

$$^1SIC(MG_2) = \ ^2SIC(MG_2) = 0.7421 \text{ bits}$$

$$^1CIC(MG_2) = \ ^2CIC(MG_2) = 0.6667 \text{ bits}$$

$$^1BIC(MG_2) = \ ^2BIC(MG_2) = 0.5775 \text{ bits}$$

As seen from the comparison with Subsection 3.B.i, the
information indices of Sarkar and Roy for some simple
multigraphs will coincide with the information on the
vertex distribution over the multigraph orbits. For more
complicated cases the vertex co-ordinates for first and
higher neighbours are not sufficient to arrive to the
graph orbits but being readily calculated they provide
practical information indices. ^1IC, ^1SIC, ^1CIC, and ^1BIC -
indices have been applied with good results to QSAR
(Quantitative Structure-Activity Relationshiphs) studies
by Basak et al. (1979, 1982) Raychaudhury et al. (1980),
Ray et al. (1981 a,b,c, 1982 a,b), and Roy et al. (1982).

J. Comparative Tables for Some Topological Information Indices

The values of six topological information indices are given as a supplement in TABLES 10,11 for some acyclic and monocyclic structures (FIG.10,22,23). More data can be found elsewhere (Bonchev et al., 1981 b; Balaban et al. 1982).

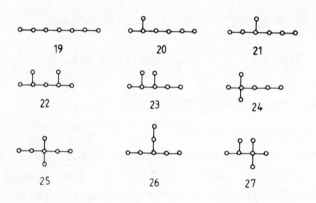

FIG.22. Acyclic graphs with seven vertices

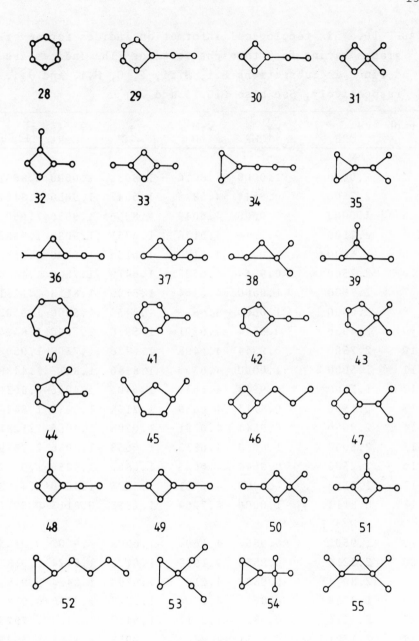

FIG.23. Monocyclic graphs with six and seven vertices

TABLE 10. Six topological information indices for acyclic
graphs having four to eight vertices (The indices are
defined in Subsections B.i, B.ii, F, G, H.i, and J.i,
respectively; See also FIG.13 and 22)

graph	$^v\bar{I}_{ORB}$	$^v\bar{I}_{CHR}$	$^v\bar{I}_D^M$	$\bar{I}_Z$	$^v\bar{I}_{C,R}$	$\bar{I}_{EDGE}^E$
	$N = 8$					
1	2.0000	1.0000	4.8016	1.8088	2.0000	0.8631
2	2.7500	0.9544	4.5870	1.6647	1.9056	1.8424
3	3.0000	1.0000	4.6049	1.8266	1.9056	1.9502
4	2.2500	0.9544	4.6120	1.6751	1.9056	1.9502
5	1.5000	1.0000	4.6120	1.6136	1.5000	1.3788
6	2.7500	0.9544	4.6322	1.6479	1.7500	1.4488
7	2.7500	1.0000	4.6364	1.6709	1.8113	2.1281
8	2.0000	1.0000	4.6515	1.8439	1.5000	1.9502
9	2.4056	0.9544	4.6201	1.5920	1.7500	1.8424
10	2.7500	0.9544	4.6498	1.6726	1.7500	1.9502
11	2.5000	1.0000	4.6352	1.8286	1.8113	1.4488
12	1.7500	0.9544	4.6679	1.6403	1.4056	0.8631
13	2.1556	0.8113	4.6539	1.2108	1.2988	1.8424
14	2.4056	0.9544	4.6751	1.6203	1.4056	2.1281
15	2.5000	1.0000	4.6833	1.6668	1.4056	2.2359
16	2.2500	0.9544	4.6625	1.6856	1.4056	1.9502
17	1.8113	1.0000	4.6751	1.8673	1.4056	1.4488
18	0.8113	1.0000	4.7064	1.2533	0.8113	0.5917
	$N = 7$					
19	1.9502	0.9852	4.1690	1.6909	1.9502	0.9183
20	2.5216	0.9852	4.1958	1.6122	1.8424	1.9183
21	2.8074	0.9852	4.2102	1.6798	1.8424	1.9183
22	1.3788	0.8631	4.2304	1.2729	1.3788	0.9183
23	2.5216	0.9852	4.2512	1.6457	1.4488	1.7925
24	2.1281	0.8631	4.2405	1.2958	1.3788	1.7925
25	1.9502	0.9852	4.2656	1.6774	1.3788	1.5850
26	1.4488	0.9852	4.2406	1.7200	1.4488	1.0000
27	1.8424	0.9852	4.2845	1.3143	1.3788	1.4592

TABLE 11. Six topological information indices for monocyclic graphs having six and seven vertices

graphs[a]	$^v\bar{I}_{ORB}$	$^v\bar{I}_{CHR}$	$^v\bar{I}_D^M$	$\bar{I}_Z$	$^v\bar{I}_{C,R}$	$\bar{I}_{EDGE}^E$
28	0.0000	1.0000	3.7821	1.6122	0.0000	0.0000
29	1.9183	1.4592	3.7962	1.5306	1.4592	1.4592
30	2.2516	1.0000	3.7454	1.6457	1.4592	1.4592
31	1.2516	0.9183	3.7962	1.3143	1.2516	1.5850
32	1.5850	1.0000	3.7821	1.5628	0.9183	1.9183
33	1.5850	0.9183	3.7534	1.2958	1.5850	0.9183
34	2.2516	1.4592	3.7198	1.5306	1.4592	0.4592
35	1.9183	1.4592	3.7710	1.2958	0.9183	1.9183
36	2.5850	1.4592	3.7454	1.5628	1.4592	1.7925
37	2.2516	1.4592	3.7821	1.5917	1.2516	1.7925
38	1.2516	1.2516	3.7962	1.3250	1.2516	2.2516
39	1.0000	1.4592	3.7821	1.5917	1.0000	1.0000
40	0.0000	1.4488	4.2665	1.6647	0.0000	0.0000
41	2.2359	0.9852	4.2548	1.6479	1.8424	1.3788
42	2.2359	1.4488	4.2398	1.6709	1.4488	1.4488
43	1.9502	1.3788	4.2849	1.5492	1.3788	1.5566
44	1.9502	1.3788	4.2776	1.6403	1.3788	1.9502
45	1.9502	1.4488	4.2595	1.5920	1.3788	1.3788
46	2.5216	0.9852	4.1968	1.6709	1.8424	1.4488
47	2.2359	0.9852	4.2373	1.6403	1.4488	1.9502
48	2.8074	1.9852	4.2373	1.6726	1.4488	2.1281
49	2.5216	0.9852	4.2055	1.6403	1.8424	1.1488
50	2.2359	0.8631	4.2432	1.2326	1.5566	1.9502
51	2.2359	0.9852	4.2548	1.6203	1.4488	1.5566
52	2.5216	1.4488	4.1775	1.2632	1.8424	1.4488
53	1.9502	1.4488	4.2512	1.6968	1.3788	1.3788
54	1.8424	1.3788	4.2714	1.2326	1.3788	1.8424
55	1.3788	1.1488	4.2849	1.2718	1.3788	1.3788

[a]Numbers correspond to the graphs given in FIG. 23

4. INFORMATION ON MOLECULAR SYMMETRY

A. A DEFINITION AND PROPERTIES

Each molecule belongs to a certain symmetry point group
(e.g.Wigner, 1931; Hochstrasser, 1966; Cotton, 1971).
The symmetry point group includes a set of symmetry operations: identity, proper rotation, reflection, inversion,
and improper rotation. A symmetry operation of a given
system is called a transformation, which leaves the whole
system in a position equivalent or identical to the initial one.

As an illustration the trans-butadiene molecule is taken into consideration below (FIG. 24):

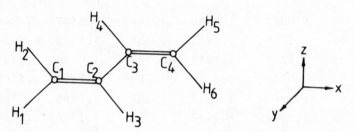

FIG.24. Structure of the butadiene molecule

This molecule belongs to C_{2h} symmetry group and has as
nontrivial elements of symmetry a plane of reflection σ_{xy},
a twofold rotation axis C_2^z, and an inversion centre i.
The symmetry operations interchange the following pairs
of atoms:

$$(C_1,C_4), \ (C_2,C_3), \ (H_1,H_6), \ (H_2,H_5), \text{ and } (H_3,H_4).$$

Thus, the distribution on symmetry of all the ten atoms
in the molecule is $P_{sum} = (2,2,2,2,2)$.

Information on molecular symmetry, I_{sym}, has been defined by Bonchev et al. (1976 c) on the basis of the atom distribution into classes of symmetry equivalent atoms. The atoms are termed equivalent if they interchange their position by the operations of the symmetry point group to which the molecule belongs.

The information on molecular symmetry complements the topological information indices discussed in Section 3. It is assumed that molecular properties are to a large extent determined by the topology of molecules. The topological description of molecules is, however, not complete, because of the failure to consider the specific molecular geometry (bond lengths and bond angles).

It should be noted that, a provision for the possiblity of defining the equivalence of points is provided in the papers of Trucco (1956 a,b) not only on the basis of the orbits of the automorphism group of a graph, but also on the basis of "each group which permutates the points of the graph". The point groups of symmetry, which consist of definite type point permutations, may also find place in this enlarged definition. The information on symmetry should be however separately defined as a special kind of information content. The very important role which symmetry plays in nature and especially in quantum mechanical interpretation of atomic and molecular states, motions, spectra, chemical properties, etc., is regarded as a sufficient reason for this definition.

It is important to take into account that the graphs do not have a fixed spatial geometry since they may be subjected to any deformations, preserving adjacency. A graph then, can be represented with a great number of variants having different symmetry. This can be used in principle to distinguish between different molecular conformations by means of their information on symmetry.

The influence of symmetry can be examined separately in molecules having equal numbers of atoms (Table 12). The molecules of highest symmetry T_d and D_{4h} (having an order of the symmetry group of h = 24 and h = 16, respectively), have the lowest amount of information, and the most asymmetric molecules of symmetry C_1 and C_s (having order h = 1 and h = 2, respectively) have the largest quantity of information;inequalites are obtained which are in good accord with the hierarchy of the groups of symmetry cited by Jaffe and Orchin (1965):

$$n = 5: \quad I_{T_d}, \ I_{D_{4h}} < I_{C_{3h}} < I_{C_{2v}} < I_{C_s} < I_{C_1}$$

$$n = 6: \quad I_{D_{2h}}, I_{D_{2d}} < I_{C_{4v}} < I_{D_{3h}} < I_{C_{2v}}, \ I_{C_{2h}}, \ I_{C_2} < I_{C_s}$$

$$n = 8: \quad I_{D_{3h}}, \ I_{D_{3d}} < I_{D_{2h}} < I_{C_3} < I_{C_2}, \ I_{C_i} \quad < I_{C_1}$$

As shown above the information index on molecular symmetry increases for equal number of atoms with increasing asymmetry. This result is consistent with the Ashby concept of information as a measure of variety in a system, since the high symmetry indicates a high uniformity as well.

B. COMPARISON BETWEEN INFORMATION ON SYMMETRY, TOPOLOGY AND ATOMIC COMPOSITION

A comparison has been made between the information indices on symmetry, topology (orbital one), and atomic composition. The three quantities are defined on molecular structures

TABLE 12. Symmetry and Information Content of Isoatomic
Molecules

No	Compound	Group of symmetry	I_{sym}, bits/ molecule
	n = 5		
1	$PtCl_4^{2-}$	D_{4h}	3.611
2	CH_4	T_d	3.611
3	CH_3Cl	C_{3v}	6.857
4	CH_2F_2	C_{2v}	7.611
5	CH_2FBr	C_s	9.611
6	CHFClBr	C_1	11.611
	n = 6		
1	C_2H_4 (planar)	D_{2h}	5.512
2	C_2H_4 (perpendicular)	D_{2d}	5.512
3	$XeOF_4$	C_{4v}	7.512
4	PF_2Cl_3	D_{3h}	8.758
5	$C_2H_2Cl_2$ (cis)	C_{2v}	9.512
6	$C_2H_2Cl_2$ (trans)	C_{2h}	9.512
7	N_2H_4	C_2	9.512
8	C_2H_3X	C_8	13.512
	n = 8		
1	C_2H_6 (eclipsed)	D_{3h}	6.488
2	C_2H_6 (staggered)	D_{3d}	6.488
3	Al_2Br_6	D_{2h}	12.000
4	CH_3CCl_3	C_3	14.490
5	$(CH_2Cl)_2$ (twisted)	C_2	16.000
6	$(CHFCl)_2$ (staggered)	C_i	16.000
7	C_2H_5Cl	C_1	24.000

in which the atoms are partitioned into different sets.
The atoms in each set are equivalent since they can interchange by means of transformations forming a group. In
the cases of topological information and the information
on atomic composition, these are permutation groups of
the vertices of the graph associated with the molecule.
At this the one-to-one mapping of the given structure onto
itself may be done by the unique demand of preserving
adjacency in the graph and chemical nature of its vertices,
respectively. The permutations of atoms in the groups of
symmetry represent definite operations of point symmetry.

The symmetry group can be presented as a permutation
group with the smallest possible order, both groups having
the same sets of equivalent atoms. Let us denote the three
permutation groups as G_{top}, G_{AC}, and G_{sym}. Since atoms of
equivalent chemical nature can differ in symmetry, then
$G_{sym} \leqslant G_{AC}$. A similar inequality $G_{sym} \leqslant G_{top}$ could be
based on the property that topologically equivalent atoms
can be chemically distinguishable (in the topological
method the vertices of the graph associated with the
molecule are considered indistinguishable), as well as
because not every inner automorphism of G_{top} represents a
symmetry operation in G_{sym}. Then at least one of the
orbits of G_{top} will be a union of G_{sym} and according to
Lemma 2.5 of Mowshowitz (1968 b) the following inequalities will hold:

$$I_{sym} \geqslant I_{AC} \tag{157}$$

$$I_{sym} \geqslant I_{top} \tag{158}$$

The information on symmetry will be equal to the information on atomic composition only when the symmetry group

contains a number of symmetry operations sufficient for treating as equivalent all atoms with the same chemical properties. This case occurs usually in molecules with a small number of atoms and high symmetry, such as methane, CH_4; ethylene, C_2H_4; acetylene, C_2H_2; methanal, CH_2O; benzene, C_6H_6, etc.

The information on symmetry and topology can be equal when each different valency in an examined molecule is realized by atoms of only one chemical element. One frequently meets this situation in organic chemistry, especially for hydrocarbons and some of the O-and N-containing derivatives. For example, for C_2H_2 and C_6H_6, the fourvalent carbons and monovalent hydrogens correspond to points of fourth and first order respectively, in the undirected graphs of the molecules, therefore, these atoms are distinguishable both chemically and topologically.

Another case of equality in (158) can occur when an isomorphic or homomorphic correspondence exists between the automorphism group of the graph and the point group of symmetry of the molecule. In both cases it is necessary for the two groups to have the same system of orbits. This can happen when G_{sym} and G_{top}^{min}, the smallest subgroups of the two groups having the same orbits as their groups, are isomorphic (e.g. the ammonia molecule). This last requirement in fact means that every automorphism from G_{top}^{min} can be presented as a symmetry operation of the molecular point group. If G_{sym} and G_{top}^{min} are not isomorphic, then some automorphisms from G_{top}^{min} are not symmetry operations acting on the whole molecule, and $I_{sym} > I_{top}$. These automorphisms can be considered as corresponding to some local symmetry operation or simultaneously to more such local symmetry operations in different parts of the molecule. "Local" symmetry here is understood as symmetry which does not encompass the entire molecule, but only part of it. Different parts of the molecule may correspond to different

164

local symmetries.

This interpretation of the automorphisms, to which no general symmetry operation of the molecular point group corresponds, finds application in the case where this type of automorphism generates greater equivalence of the atoms.

As an illustration of the last case, let us examine the propylene molecule, C_3H_6 (FIG.25). Automorphisms are possible here which make the hydrogen atoms {4,5} and {7,8,9} equivalent.

Expressed in cyclic notation, these are the transformations (4,5), (7,8), (7,9), (8,9), (7,8,9), and (7,8,9) and some of their combinations. Local symmetry operations (FIG.25) correspond to each of these cyclic permutations. The group $=CH_2$, including atoms 4,5, has C_{2v} local symmetry with $(E, \sigma_{xz}, \sigma_{xy}, C_2^x)$ as the symmetry operations. For

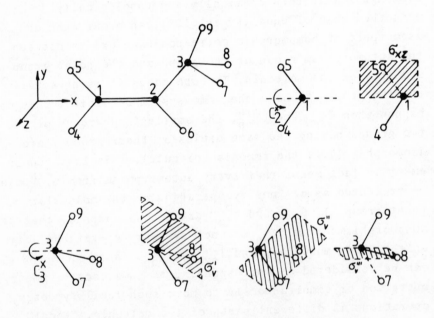

FIG.25. Local symmetry in the propylene molecule

atoms 7, 8, and 9, with C_{3v} local symmetry of the $-CH_3$ group, these are C_3, C_3^2, σ_v', σ_v'', and σ_v'''.

However, the total symmetry of propylene is much lower than the C_{2v} and C_{3v} local symmetries of its parts. It is either C_1, i.e. total asymmetry, when the methyl group $-CH_3$ is oriented arbitrarily in space, or C_s, i.e. with a unique plane of reflection σ_{xy}, when one of the three hydrogen atoms lies in the σ_{xy} plane, and the other two are symmetrically situated above and below this plane. In the case of C_1 symmetry there are no equivalent atoms in the molecule at all, while when C_s symmetry applies, only two of the three hydrogen atoms of the methyl group are equivalent. In the propylene molecule, therefore, symmetry generates fewer equivalent atoms than topology, the information on symmetry will be greater than the topological information.

The two different values for I_{sym} of the propylene molecule correspond to different spatial orientations of the methyl group, i.e. to different conformations of this molecule. This example indicates that the symmetry method allows, in principle, a different information content to be derived for the different conformations of a given molecule, if these differ by symmetry. This may be of importance especially in organic chemistry, where the chemical structures often exist in various conformations with different physical and chemical properties.

The analysis made above indicates that the differences between the information on symmetry and on topology are effected by two factors: the failure to consider the differences in the chemical nature of the atoms in the topological method, and the presence of inner automorphisms, which do not correspond to a total symmetry operation, but only to a local symmetry operation (or combination of such local operations). Hence a conclusion may be drawn concerning the possibility of bringing together these two information methods. This could occur, initially, if the physical differentiation of the points in the graph is introduced preliminarily in the topological method (i.e. if the topological differences only of points of the same chemical nature are taken into account), and secondly, if the local symmetry in separate parts of the molecule (and above all in atomic groups with simple structure such as $-CH_3$, $-CH_2$, $-NH_2$, $-NO_2$) is used additionally to the molecular symmetry method. In essence it means proposing a method which is a hybrid between the topological method and the molecular symmetry method. In that case, the two methods modify themselves in such a way so as to define one and the same information content, $I'_{sym} = I'_{top}$.

It should be noted that, in this combined method, the idea of equivalent atoms in molecules corresponds, in some cases, more to their chemical behaviour than when defined according to the initial methods. These are cases when the equivalence of a given group of atoms (for example hydrogens in $-CH_3$, $-NH_2$, and other groups) cannot be provided by a total symmetry operation in the molecule, but is actually ensured by the presence of free rotation of the whole atomic group around a single chemical bond.

Up to this point, the information on symmetry and the topological information were compared in the case where only an undirected graph was used for a given molecule. However a great number of directed graphs with different information content may correspond to a given molecular structure. Mowshowitz (1968 b) proved that every possible decomposition of a finite graph having n points to different orbits could be realized by a suitable orientation of the graph vertices. Therefore, different directed graphs, representing a molecular structure of n atoms, will have an information content which includes each of the possible values in the interval $0 \leqslant I_{top} \leqslant$ lb n. One can generalize that, in the case of directed graphs, the information on symmetry of a given molecule could not only be greater than or equal to the topological information (as in the case of undirected graphs, eq 158), but could be also smaller than it.

This indicates that among the large number of directed graphs, there will always be one whose topological information content is equal to the information on symmetry of the molecular structure examined. The question of the possibility of finding in advance this directed graph for an arbitrary molecule remains open. It is associated with the problem of a simple orientation of the graph edges, so that the same set of equivalent atoms is to be obtained as by the effect of the symmetry operations. As a consequence of the Mowshowitz work (1968 c), however, the inverse problem may be considered as solvable. After a determination of the information on symmetry of a given structure, a directed graph can be constructed having the same information content.

C. INFORMATION ON SYMMETRY, TOPOLOGY AND ATOMIC COMPOSITION OF SOME GROUPS OF CHEMICAL COMPOUNDS

The numerical values of the three kinds of information indices, compared in Subsection B, are presented in TABLE 13 for some organic compounds. More data can be found elsewhere (Bonchev et al., 1976 c).

TABLE 13. Information Content of Some Organic Compounds[*]

No	Compound	Group of Symmetry	I_{cc}	I_{top}	I_{sym}
	Alkanes				
1.	Methane	T_d	3.611	3.611	3.611
2.	Ethane	D_{3h}	6.488	6.488	6.488
3.	Propane	C_{2v}	9.303	18.55	24.06
4.	n-Butane	C_{2h}	12.09	25.79	31.31
5.	n-Pentane	C_{2v}	14.86	39.98	45.49
6.	n-Hexane	C_{2h}	17.63	48.93	54.45
7.	n-Heptane	C_{2v}	20.39	64.54	70.05
8.	n-Octane	C_{2h}	23.15	74.71	80.22
9.	n-Nonane	C_{2v}	25.91	91.38	96.89
10.	n-Decane	C_{2h}	28.66	102.5	108.0
	Aromatic Hydrocarbons				
11.	Benzene	D_{6h}	12.00	12.00	12.00
12.	Naphthalene	D_{2h}	17.85	41.07	41.07
13.	Anthracene	D_{2h}	23.52	66.05	66.05
14.	Phenanthrene	C_{2h}	23.52	86.05	86.05
15.	Pyrene	D_{2h}	25.00	76.22	76.22
	Other Compounds				
16.	Quinone	D_{2h}	17.51	23.02	23.02
17.	Adenine	C_1	23.78	56.61	58.61
18.	Thymine	C_1	27.49	53.86	58.61
19.	Guanine	C_1	29.17	62.00	64.00
20.	Cytosine	C_1	23.75	46.11	48.11

[*]in bits per molecule

The information index on molecular symmetry has been applied to correlations with various properties of the chemical compounds such as melting points, vapour pressure, surface tension, entropy, etc. (Bonchev et al, 1976 c; Bonchev and Mekenyan, 1973).

The first steps towards another information theoretic approach to molecular symmetry have also been made. Subject of consideration in this approach are directly the symmetry groups regarded as mathematical stuctures, composed of classes of symmetry elements, irreducible representations and their characters, subgroups, etc. (Bonchev and Lickomannov, 1977 b). The impact, which the statistical properties of the point groups thus defined could have on particular molecular systems, would be studied as a second stage in this approach.

5. INFORMATION ON MOLECULAR CONFORMATIONS

Structures generated by means of a free rotation of atoms or groups of atoms about single chemical bonds are called *conformations*. They are also called *rotational isomers* (rotamers) due to their interconversion without bond breaking.

The number of conformations for a certain molecule is the source of another kind of structural information contained in it. Regarding different conformations, N_{CONF} in number, as equiprobable the mean *information on molecular conformation*, $\bar{I}_{CONF}$, has been defined (Zhdanov, 1967):

$$\bar{I}_{CONF} = 1b \ N_{CONF} \qquad (159)$$

This information index has been used in the definition
of the so-called total active information capacity of a
molecule (Section V.4).

By determining a general index of molecular complexity
(Section V.3), Dosmorov (1982) made use of the equation
for the total information on molecular conformation,
$\bar{I}_{CONF}$:

$$I_{CONF} = N_{CONF} \text{ lb } N_{CONF} \tag{160}$$

The simplest example is the cyclohexane molecule which
is known to exist in two conformations, called a chair
and a tub. Hence, $N_{CONF} = 2$, $\bar{I}_{CONF} = 1$ bit per conformation, $I_{CONF} = 2$ bits per molecule.

6. INFORMATION ON MOLECULAR COMPOSITION

Dealing with complicated molecules like DNA, proteins,
polymers, etc. atoms are no more regarded as convenient
structural elements and atomic groups or molecular fragments come into consideration. Such an approach has been
developed in connection with the study on the information
properties of some biologically important structures.

Branson (Quastler, 1953) calculated the *information on
aminoacid composition,* I_{AAC}, of 26 proteins. Proceeding
from the negentropy principle of information (Chapter I)
he applied the formula:

$$I_{AAC} = k(\ln N! - \sum_{i=1}^{k} \ln N_i!) \tag{161}$$

where N_i is the number of aminoacid residues of type i, N is their total number, and k is the Boltzmann constant.

Thus for the insulin molecule (N = 103) the number of aminoacid residues is as follows: glycine (7), alanine (6), valine (8), leucine (12), iso-leucine (3), penylalanine (6), proline (3), serine (6), threonine (2), tyrosine (9), cystine (12), asparaginic acid (6), glutamic acid (15), hystidine (4), arginine (2), and lysine (2).

Applying eq. (161), $I_{AAC} = 252.7$ k is obtained. Dividing by N, the mean amount of information per an aminoacid residue is obtained to be 2.45 k.

As a general feature of all the proteins present in organisms, the Branson information index has been found to exceed 70% of its maximum value. This is supposed to indicate a connection between the information content and functions of protein molecules.

7. INFORMATION ON CONFIGURATIONS

The information on aminoacid composition, defined in Section 6, is calculated without taking into account the mutual influence of the individual aminoacids. The latter is accounted for in the definition of the *information on configurations*, I_{CFG}, of polypeptide chain which is added by Augenstine (Quastler, 1953) to I_{AAC} to calculate the total structural information content of proteins.

The exact calculations of the configurational information are hindered by the enormous number of configurations, as well as by the various types of intramolecular bonding decreasing this number. Due to this, estimations of the upper and low bounds of this index are only given for helical structures:

$$\bar{I}_{CFG}(\text{low bound}) = (3 \text{ lb } N)/N, \text{ bits per residue} \qquad (162)$$

$$\bar{I}_{CFG}(\text{upper bound}) = 0.50 + (30 + 3 \text{ lb}N)/N, \text{ bits per residue} \qquad (163)$$

where N is the total number of the aminoacid residues.
The informations on configurations and molecular composition defined above have been applied to the estimation of the amount of information necessary for the normal activity of a protein molecule.

8. ELECTRONIC INFORMATION INDICES

A. ELECTROPY AND BONDTROPY INDICES

The information indices presented in Sections 1 to 7 have the general feature of dealing with molecules as entities composed of atoms (or atomic groups) and bonds. This is largely a macroscopic viewpoint which does not take into account the electronic factor. The latter is, however, of crucial importance in case of heteroatomic molecules. Thus, information indices based on the electronic structure of molecules had to come into consideration.

The first information index dealing with electronic distributions in molecules has been introduced by Yee et al. (1976). Electrons or bond pairs are regarded as distributed into partial bond spaces while the whole isolated molecule is assumed to form a finite space, S. The latter is divided first into two basic types of partial spaces: the valence bond space S_v, comprising the electrons that contribute to chemical bonding and the nonvalence space S_n, regarded as a core part. S_v is further divided into σ - and π - partial bond spaces. In terms of the set theory, the partition of the total molecular space is expressed by eq (164):

$$S = S_n \cup S_v = S_n \cup S_\sigma \cup S_\pi \tag{164}$$

The σ - bond space can be divided further into various σ - partial bond spaces, such as S_{CC}, S_{CH}, C_{CH_2}, C_{CH_3}, etc., proceeding from the concept of chemical bonds.

It should be stated here that dealing with the π - bond space as one partial bond space for the whole molecule is justified for molecules with a total π - electron system (or otherwise, conjugated π - bonds). When molecules having isolated π - bond or π-fragments come into consideration S_π have to be also divided into individual contributions in the same manner as for S_σ.

Three methods of treating partial bond spaces S_n, S_{CH}, S_{CH_2}, and S_{CH_3} have been examined:

(i) *United model:* All partial bond spaces of the same kind are assumed to form a common bond space. For example, all 1s-electrons of the second row chemical elements are assumed to form one partial space (S_n).

(ii) *Separated model:* Each partial bond space, no matter whether it be of the same kind or not, is assumed to form an individual bond space. For example, the two methyl groups in the ethane molecule belong to different bond spaces.

(iii) *Semi-separated model:* The partial bond spaces S_v are treated in the same way as in (i), while the partial bond spaces S_n as in (ii).

174

The above three models, however, treat in the same way the partial bond space S_{CC} formed from the carbon skeleton of a molecule. It is assumed to be one partial space when no heteroatoms appear in it. In the opposite case, when the carbon skeleton is fragmented by any heteroatom, as in the case of heteroaromatic compounds, ketones, secondary and tertiary amines, S_{CC} is partitioned into several subspaces.

The three models are exemplified in TABLE 14 by the normal pentane molecule.

Once the partial bond spaces are uniquely defined, the electrons or bond pairs are regarded as particles of a microcanonical ensemble. Distribute N particles into k partial bond spaces of S, i.e. N_1 particles into partial bond space S_1, N_2 particles into partial bond space S_2, etc. The total number P of the possible different ways of

TABLE 14. Number of electrons and bond pairs in the partial bond spaces of normal pentane

Model	Partial bond spaces for electrons					Partial bond spaces for bond pairs				
	1s(C)	C-C	CH	CH_2	CH_3	1s(C)	C-C	CH	CH_2	CH_3
i	10	8	0	12	12	5	4	0	6	6
ii*	2x(5)	8x(1)	0	4x(3)	6x(2)	1x(5)	4x(1)	0	2x(3)	3x(2)
iii*	2x(5)	8	0	12	12	1x(5)	4	0	6	6

*Figures in parentheses are the numbers of bond spaces defined

distribution is

$$P = N! / \prod_{i=1}^{k} N_i!$$ (165)

where

$$\sum_{i=1}^{k} N_i = N$$ (166)

Taking a logarithm of P at base two a new information index called *electropy* is defined:

$$\varepsilon = lb\ P = lb\ (N! / \prod_{i=1}^{k} N_i!)$$ (167)

For both N, $N_i \gg 1$ eq (167) transforms into eq (7) for the total information content of the molecule.

When bond pairs instead of electrons are regarded as distributed into the partial bond spaces of the molecule, a similar index called *bondtropy* is defined.

The electropy values for normal pentane are calculated from eq (167) as follows:

Model i: $\varepsilon = lb(42!/(10!8!12!12!)) = 79.7327$ bits

Model ii: $\varepsilon = lb(42!/((2!)^5(8!)^1(4!)^3(6!)^2)) = 126.0410$ bits

Model iii: $\varepsilon = lb(42!/((2!)^5 8!12!12!)) = 91.9388$ bits

Similarly, the bondtropy values are: $\varepsilon = 34.9941$ bits (Model i), $\varepsilon = 52.7148$ bits (Model ii), and $\varepsilon = 41.9010$ bits (Model iii).

Applying electropy to correlations with some thermodynamic properties of hydrocarbons, Model (ii) has been found to be the sole model satisfying the experimental data.

The physical meaning of the electropy index is given as the degree of freedom in choosing electrons (or bond pairs) to be located in various partial bond spaces of a molecule in the process of molecular formation.

A modification of the electropy index, ε', has also been proposed in which the steric hindrance has been taken into account. These steric effects which largely influence the differences in the properties of isomeric compounds have been characterized by means of an additive steric parameter ρ:

$$\varepsilon' = \varepsilon - lb\rho = \varepsilon - \Sigma \ lb \ \rho_i \qquad (168)$$

Here $\rho_i = 1$ when hindrance is lacking; $\rho_i = 2$ when steric hindrance contributes to one pair of the adjacent methyl groups, linked to one carbon atom; $\rho_i = 6$ and $\rho_i = 12$ for the three and four methyl groups linked to one carbon atom. For instance $\rho = 1$, for 3- methyl pentane, while $\rho = 2$ for 2- methyl pentane. The modified electropy values have been calculated for a number of alkanes, cycloalkanes and monoolefins and applied to correlation with various properties of these compounds (Sakamoto et al., 1977).

Yee et al. (1976) have also specified another modification of the electropy index making use of another information function called *redundancy of information*. As elicited by information theory, each message is usually transmitted by means of more symbols than the minimum number needed so as to compensate the possible errors during transmission. The redundancy of electropy $R(\varepsilon)$ is similarly defined:

$$R(\epsilon) = 1 - \epsilon/\epsilon_{max} \tag{169}$$

where ϵ_{max} is the possible maximum electropy value for an ensemble of isomeric molecules (e.g. normal alkane in the case of paraffin isomers).

$R(\epsilon)$ has been interpreted as the measure of restricting the degree of freedom in distributing a fixed number of electrons into any molecular partial bond space. The two extremes, $R(\epsilon) = 0$ and $R(\epsilon) = 1$, correspond to the maximum and minimum degrees of freedom, respectively. Specifically, $R(\epsilon) = 1$ occurs when there is a single way of distributing the electrons. The redundancy of electropy is regarded as an inverse proportional measure of the molecule bulkiness. It can also be applied in correlations with isomeric molecules.

The electropy approach offers the essential advantage of being applicable to heteroatomic molecules. It can also be of use as a measure of biomolecule complexity.

B. INFORMATION INDEX OF ELECTRON DELOCALIZATION (AROMATICITY)

The electropy index discussed in Subsection A deals with electrons in molecules localized in core or bond pairs. Another information approach has been recently proposed (Fratev et al., 1980) to describe electron delocalization in molecules, and more specifically in aromatic compounds.

The electron distribution in molecules can be described by means of the reduced density matrix (i.e. the one-particle matrix) which in the LCAO - approximation is presented by the so-called charge - bond order matrix. Within some LCAO methods like EHT, PPP, CNDO, etc., it is possible

to define an analogous matrix for the electron energy
distribution over AO's or among them (E.g.McWeeny and
Sutcliffe, 1969). These quantumchemical matrices - $Q(Q_{11}$,
$Q_{12}, \ldots, Q_{\mu\nu}, \ldots)$ give information on the character of
the distribution of the matter (electrons) or energy in a
molecule. If the elements $Q_{\mu\nu}$ are transformed:

$$\frac{Q_{\mu\nu}}{A} = p_i, \text{ where } A = \Sigma \, Q_{\mu\nu} \tag{170}$$

then the quantities $\{p_i\}$ ($i \in < 1$, $k >$ and $\Sigma \, p_i = 1$) can be
regarded as relative characteristics of the electron (or
energy) distribution over and between AO's, which for each
matrix (i.e. for each M or M_i in the electron state α) is
normalized to unity. Moreover, the distribution has a
probabilistic character since the elements (i.e. the
charges and bond orders) reflect the probability of the
electron distribution over and among AO's. The finite
probability scheme $P\{P_1, P_2, \ldots, P_k\}$ is constructed and
its entropy $H(P)$ is calculated by means of the Shannon
equation (2). The maximum of entropy corresponds to the
uniform distribution $P_1 = P_2 = \ldots = P_k$:

$$H_{max}(P) = lb \, k \tag{171}$$

The deviation of entropy of the electron distribution
from its maximum value is proposed as a new information
index:

$$I = H_{max}(P) - H(P) \tag{172}$$

Replacing eq (2, 170, 171) into (172) the information index is obtained in its final form:

$$I(\alpha) = \mathrm{lb}\ (\frac{k}{A}) + \frac{1}{A} \sum_{\mu\nu} Q_{\mu\nu}\ (\alpha)\ \mathrm{lb}\ Q_{\mu\nu}\ (\alpha),\ \text{bits} \qquad (173)$$

Here $I(\alpha)$ is defined as a measure of the nonuniformity in the distribution of electrons (the entries of the charge - bond order matrix $P_{\mu\nu}$ or energy $E_{\mu\nu}$ - matrix). $P_{\mu\nu}$ - matrices can be calculated within each LCAO-MO-Cl method for an arbitrary electron state α, while $E_{\mu\nu}$ - only under the restrictions of ZDO and EHT - approximations (See e.g. McWeeny and Sutcliffe, 1969). The uniformity in the electron distribution is, however, an essential feature of aromatic (delocalized) molecules. Benzene is the best example having completely equalized bonds and atomic charges. Hence, $I(\alpha)$ can be viewed as an information index which is the reciprocal measure of aromaticity (delocalization).

This index has been normalized, using benzene as a reference molecule, so as to make it increase with aromaticity within the (0,1) range:

$$\tilde{I}(\alpha) = I^B(0)/I^{M(\text{or } M_i)}\ (\alpha) \qquad (174)$$

Here $I^B(0)$ is the reference index of benzene in the ground state ($\alpha = S_o$, $Q_{\mu\nu} = P_{\mu\nu}$ or $E_{\mu\nu}$). $I^M_i\ (\alpha)$ is the information index of the molecular fragment M_i in the electronic state α. It is also calculated by eq (173) but the summation is over $(\mu,\nu) \in M_i$ only.

The information aromaticity index proved to be a convenient tool for quantitative estimates of ring and total aromaticity of molecules in their ground and excited states.

C. INFORMATION ON THE NODAL PROPERTIES OF Π-MOLECULAR ORBITALS

Another information theoretic approach to delocalized
molecular orbitals has been developed by Bonchev and Lickomannov (1979). A set of molecular spatial regions between every pair of adjacent atoms in the molecule is
considered. These regions may have or have not a node,
manifesting nodal or non-nodal, properties, respectively
(FIG.26 a). When the node is not located between two
adjacent atoms, but at a given atomic nucleus (FIG.26 b),
half of the interatomic regions of this kind are nodal
(that which borders the node), and the other half are non-
nodal. The set of Π-MO spatial regions thus constructed

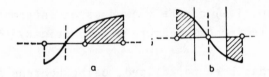

FIG.26. Partitioning of the Π-molecular orbital space
into nodal (⬜) and non-nodal (▨) regions

(m in number) has n nodal and $m-n$ non-nodal regions.
Applying eq (7) the information index on the nodal properties of Π-MO is specified in bits per orbital.

$$I^{NP} = m \; lb \; m - n \; lb \; n - (m - n) \; lb \; (m - n) \qquad (175)$$

Benzene may be taken as an example illustrating the
procedure. The nodal properties of the molecular orbitals
of benzene are depicted in FIG. 27.
In annulenes the cardinality n of the set constructed
from the nodal and non-nodal regions (i.e. the total

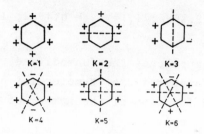

FIG.27. Nodal planes of the benzene π-MO's

number of π-MO's) is equal to the number of carbon atoms N: $n = k_{max} = N$. The following sets are formed for the benzene π-MO's, in which the number of the non-nodal and nodal regions are expressed as ordered pairs:

$$S_1(6,0); \ S_2, \ S_3(4,2); \ S_4, \ S_5(2,4); \ S_6(0,6)$$

Hence, applying eq (175) one obtains the information on the nodal properties of the benzene π-MO's to be zero for $k = 1$ and $k = 6$ while $I^{NP} = 5.5098$ bits for $k = 2$ to 5.

As seen, the symmetry that exists among the bonding and antibonding π-MO's in alternant hydrocarbons (Coulson and Rushbrooke theorem) results in the same I^{NP} values for each pair of corresponding MO's having Hueckel energies $\pm\varepsilon_k$. The information index, however, can be modified so as to distinguish between each two corresponding bonding and antibonding orbitals treating the ordered pairs S_k by the Muirhead reordering procedure (See Subsection 3.H). The reordered sets of benzene non-nodal and nodal interatomic regions are as follows:

$$S_1'(6,6); \ S_2',S_3'(4,6); \ S_4',S_5'(2,6); \ S_6'(0,6)$$

The respective values of the modified information index decrease regularly with increasing number of MO:

$$I_1 = 12; \ I_2 = I_3 = 9.7098; \ I_4 = I_5 = 6.4902; \ I_6 = 0 \text{ bits.}$$

For polyenes and annulenes the (unmodified) information index on the nodal properties of π-molecular orbitals has been expressed as a function of the number of carbon atoms N and the MO label k only:

Polyenes:

$$I_k^{NP} = (N-1) \ lb \ (N-1) - (N-k) \ lb \ (N-k) - (k-1) \ lb \ (k-1) \quad (176)$$

Annulenes:

$$I_k^{NP} = N \ lb \ N - (N-k) \ lb \ (N-k) - k \ lb \ k; \ k = \text{even} \quad (177)$$

$$I_k^{NP} = N \ lb \ N - (N-k+1) \ lb \ (N-k+1) - (k-1) \ lb \ (k-1); \quad k = \text{odd} \quad (178)$$

The information analysis of the nodal properties of the polyene and annulene π-MO reveals some interesting features. The frontier orbitals (HOMO and LUMO) have the maximum information content which in bits is equal to the number of carbon atoms N for 4n- annulenes and equals N-1 for odd polyenes. LOMO and HUMO have the minimum information content as compared with the other orbitals. This content is zero for polyenes and even annulenes.

Another interesting result is that the information on nodal properties of the frontier orbitals in most cases is conserved during the transmutation of a certain structure into another one. For instance, in the case of polyenes and annulenes a typical transmutation is given below:

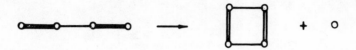

This holds for all N-membered rings except the most stable 4n+2-ones:

$$I^{N+1\text{-polyenes}}_{HOMO(LUMO)} = I^{N\text{-annulenes}}_{HOMO(LUMO)} \qquad (179)$$

where N = 4n, 4n + 1, and 4n + 3.

The maximum information content of the frontier Π-MO might be of interest to chemical theory due to the great importance of these orbitals in the studies on chemical reactivity. On the other hand, the conservation of the maximum information content for the frontier orbitals during some chemical reactions raises the question of whether the Woodward-Hoffman rules (Woodward and Hoffman, 1970; Fleming, 1976) could also be associated with the supposed general principle of maximum information as found in the case of the Pauli exclusion principle and the first Hund rule (See Chapter III).

Another possible important aspect of this information approach is related to the possibility of predicting nodal properties of Π-MO's in different series of compounds taking polyenes or annulenes as reference compounds.

Concluding this Section the studies of Bartel (1975) should be mentioned. Bartel has examined the possibility of predicting the orientation in electrophilic aromatic substitutions by comparing information indices (calculated from charge densities) and the o-, m-, and p- topological orientation probabilities.

CHAPTER 5
General Molecular
Information Indices

1. INDICES COMBINING INFORMATION ON STRUCTURE AND ATOMIC COMPOSITION

The separate molecular information indices, discussed in detail in Chapter IV, characterize various individual features of molecules. It has been recognized since the time of introducing the first information indices that a single index cannot be a general molecular characteristic. Hence the need of combined molecular information indices had been present.

The first *general molecular information indices* have been composed of a structural term, and a term for the kind of atoms in the molecule. Rashevsky (1955) combined his topological (or more precisely, orbital) information (Chapter IV, Subsection 3.B.i) with the information on the atomic composition (Chapter IV, Section 1):

$$I_{RASH} = I_{ORB} + I_{AC} \tag{180}$$

A similar approach had been developed by Morovitz (1955) who added the information on atomic composition to another structural term, the Morovitz information on the ways of linking the atoms in a molecule, I_{VB} (Section IV.2):

$$I_{MOR} = I_{VB} + I_{AC} \qquad (181)$$

Eq (180) and (181) should be viewed as the first attempts to develop a general information index of molecular complexity.

2. INDICES COMBINING INFORMATION ON MOLECULAR AND ATOMIC STRUCTURES

The total molecular information content may also include atomic contributions (Bonchev and Peev, 1973, Bonchev, 1979), i.e. it may be defined as a sum of a certain kind i of molecular information index I_{mol}^i (e.g. information on atomic composition, topology or symmetry) and the information content of all atoms, I_{at}^j:

$$I_{mol}^{total} = I_{mol}^i + \sum_k n_k I_{at,k}^j \qquad (182)$$

Here the superscript j stands for one of the kind of atomic information indices defined in Chapter III (e.g. information on the electron distribution over shells, subshells, atomic orbitals, etc.). n_k is the number of atoms of kind k, and the summation in eq (182) is over all the atoms in the molecule.

I_{mol}^i, specified in Chapter IV, are different atom or bond distributions in a molecule. The electronic structure of the molecule is not taken into consideration within this approach. Thus the additive scheme (182) is based on similar ideas with the valence bond method in quantum chemistry where the atoms are assumed to preserve their individuality. In both approaches electrons come into consideration as electrons of a certain atom. The information approach defined by means of eq (182) is called *"atoms in molecule"*. It has been applied to correlations with some properties of chemical compounds (Bonchev and Mekenyan, 1973; Bonchev et al., 1975).

Another information approach has also been proposed (Bonchev, 1979) whose ideas are to some extent close to those of the molecular orbital method in quantum chemistry. In both approaches molecules are regarded as composed of atomic nuclei and electrons whilst the atoms in molecules are viewed as having completely lost their individuality. This information approach could be termed *"nuclei and electrons"*. In it, the total molecular information content, I_{mol}^{NE}, is defined as a sum of the information on the particle distribution within the nucleus and electron shell, respectively:

$$I_{mol}^{NE} = I_{mol}^{nucl} + I_{mol}^{el} \qquad (183)$$

The various information indices on chemical composition, topology and symmetry, introduced in Chapter IV as atom distributions in a molecule, might be used as a first term in eq (183). The second term in (183) might be represented by the electropy, the information index on electron delocalization (aromaticity), etc. (See Chapter IV, Section 8). The information approach "nuclei and electrons" might be expected to develop rapidly similarly to molecular orbital method, thus contributing to the interplay between Information Theory and Quantum Mechanics.

3. GENERAL INDICES OF MOLECULAR COMPLEXITY

A. THE BERTZ INDEX

Two more general indices of molecular complexity appeared simultaneously in 1980/1981 in USSR and USA.

Searching for a convenient criterion that can reflect the increasing molecular complexity when synthesizing complex organic compounds Bertz (1980, 1981) proposed a new general information index. Similarly to the indices discussed in Section 1 the new index incorporates the information on atomic composition and a structural term, the information on the graph connections, I_{CONN} (Chapter IV, Section 3.J.ii). The latter reflects in more detail the topological structure of a molecule than the structural indices of Rashevsky (1955) and Morovitz (1955) used in eq (180) and (181). In particular, molecules with multiple bonds (or their multigraph) are readily treated in this way. Chirality and stereochemistry are also reflected by the distribution of connections into classes of orbital equivalency. In addition, a third term has been used by Bertz, to reflect properly the molecular size:

$$I_{size} = N \; lb \; N \qquad (184)$$

where N can represent any graph-theoretical invariant (vertices, edges, connections, distances, etc.). The choice of the specific invariant is governed by the chemical problem to be solved. The need of such term originates from the zero information content of systems of high symmetry where all the invariants used are equivalent and the influence of the molecular size is completely eliminated.

In the light of the foregoing, the general index of
molecular complexity of Bertz is given by the equation:

$$I_{Bertz} = I_{AC} + I_{CONN} + I_{SIZE} \tag{185}$$

B. THE DOSMOROV INDEX

A similar but more detailed approach has been developed by Dosmorov (1980; 1982) with the purpose of a priori
generation of the possible reaction mechanisms:

$$I_{Dosm} = I_{AC} + I_A + I_B + I_{SYM} + I_{CONF} \tag{186}$$

Here the information on the kind of atoms, I_{AC}, is followed by the information contributions of all atoms in
the molecule, I_A, calculated by the Bonchev and Peev (1973)
equation (See Chapter III, Section 1). Incorporating the
atomic information content, the Dosmorov index is more
discriminating than the Bertz index in case of different
substituent atoms of the same valence (e.g). halogen
substituted hydrocarbons). The fourth term, I_{SYM}, represents the information index based on the symmetry point
group (Chapter IV, Section 4). The third term has been
introduced by Dosmorov to account for the number and kind
of chemical bonds and non-bonded interactions (Chapter IV,
Section 2). The fifth term, the information on molecular
conformations has already been introduced in Chapter IV,
Section 5. In this way eq (186) specifies a completely
general index of molecular complexity. The molecular size
is indirectly taken into account in the atomic information
indices I_A. The single point to be discussed in connection
with eq (186) is whether the I_B index is the best term

representing the molecular topology. Certainly the information on molecular graph connections of Bertz, I_{CONN},
could be preferred being more closely related to molecular architecture.

In the light of the discussion in Section 2 it can be
concluded that both general indices of molecular complexity, given by eq (185) and (186), are defined within the
approach "atoms in molecules", i.e. the molecule is considered as composed of atoms, connected by bonds. The other
approach "nuclei and electrons" (eq. 183) can also be
conveniently extended so as to serve as a general molecular complexity index in which the electronic factor has
been taken into account. The extension can be done by
means of one of the indices I_{EL} introduced in Chapter IV,
Section 8. We propose below such a general molecular information index:

$$I_{MOL} = I_{IC} + I_{NUCL} + I_{TOP} + I_{SYM} + I_{EL} + I_{CONF} \qquad (187)$$

Here, considering the molecule as composed of nuclei and
electrons, the information on the kind of nuclei, or
otherwise, the information on the isotopic composition,
I_{IC}, is included instead of the information on atomic composition. Both indices coincide when each chemical element
in the molecule is represented by one isotope only, but
differ when more isotopes are available since the atoms
are distributed in equivalent classes according to the
number of protons (the atomic number) only while isotopes,
according to the number of protons z and neutrons n..
Thus, in case of HDO there are two equivalent hydrogen
atoms (Z = 1), but two different isotopes of hydrogen
(n = 1 and n = 2, respectively). Hence, the atomic composition distribution of this molecule is 3(2,1) while the
isotopic one is 3(1,1,1). I_{NUCL} is a sum over the inform-

ations on the proton-neutron composition of all the nuclei, as defined in Chapter III, Section 1:

$$I_{NUCL} = \sum_{i=1}^{k} I_{n,p}^{i} \tag{188}$$

I_{NUCL} also accounts for the molecular size by means of the number of atomic nuclei k.

The term I_{TOP} in (187) may be a representative term for molecular topology selected out of the numerous information indices introduced in Chapter IV, Section 3. None of the latter, however, could provide a *comprehensive* description of the molecular topology. For this reason the question of the proper representation of I_{TOP} may be regarded open. A more general approach to this question is discussed below in Section 4.

4. THE TOPOLOGICAL SUPERINDEX

Aiming at the constructing of a powerful index for the discrimination of isomeric molecules, a topological information superindex has been devised by Bonchev et al. (1981 b) which consists of a number of topological information indices. In the first version the superindex SI is a sum of individual indices representing different features of the molecular graph:

$$SI = {}^{v}I_{orb}^{E} + {}^{v}I_{chr}^{E} + {}^{v}I_{C}^{E} + I_{edge}^{E} + {}^{v}I_{D}^{M} + I_{Z} \tag{189}$$

Here the six terms represent the orbits, the chromatic properties, the radial centric structure, the edges (and vertex degrees), the distances, and the non-adjacent numbers of the graph. The superindex SI is, however, rather an open system and can be extended so as to represent other graph characteristics like adjacency, incidence, and cycle matrix, etc.

Developing further the idea of a superindex, the sum of information topological indices has been replaced by a sequence of such indices, due to the fact that a great deal of the information they contain is lost upon summation in eq (189). Actually, this is the major pitfall of all other general indices discussed in this chapter. The amended version of the superindex, that retains all the information contained in the individual indices, is presented in the form:

$$SI = \{{}^{v}I_{orb}^{E}, \; {}^{v}I_{chr}^{E}, \; {}^{v}I_{C}^{E}, \; I_{edge}^{E}, \; {}^{v}I_{D}^{M}, \; I_{Z}, \; ...\} \tag{190}$$

Clearly, eq (190) can be extended so as to incorporate other topological or nontopological indices like those on atomic composition, size, symmetry, conformations, etc., arriving thus to a more general index of molecular complexity.

The application of eq (190) will be discussed in Chapter VI related to the isomer discrimination problem.

5. THE ACTIVE INFORMATION CAPACITY OF MOLECULES

The information indices introduced in Chapters IV and V deal with the so-called bonded information or information which is contained in molecules. A concept of the active information capacity of organic molecules has been proposed by Zhdanov (1967; 1979). The following discrete molecular states are regarded as active in the information sense: atomic asymmetry, molecular asymmetry, cis-trans isomerism, stable conformations and tautomerism (including cycle-chain tautomerism, as well as the internal salt formation). This approach is an extension of the Rackow (1963) ideas who estimated the information capacity of an asymmetric carbon atom to be 1 bit per molecule (or 1 molbit) taking into account the equal probability of the D- and L- configurations.

The total active information capacity of a molecule is determined by summation over all individual contributions listed above:

$$I_a = \sum_{i=1}^{5} I_{a,i} = I_{as} + I_{mas} + I_{ct} + I_{CONF} + I_t \qquad (191)$$

This approach is exemplified by means of the glucose molecule:

$$
\begin{array}{c}
CHO \\
\vdash OH \\
HO \dashv \\
\vdash OH \\
\vdash OH \\
CH_2OH
\end{array}
$$

There are four asymmetric centres, and a fifth one arises upon the formation of a cyclic structure. Hence, I_{as} = 5 bits. There is also tautomerism between the open, furanosic and pyranosic structures, as well as two possible conformations for which it is obtained I_t = 1b 3, and I_{CONF} = 1 bit, respectively. The total information capacity of glucose molecule is thus calculated to be 7.585 bits

To avoid the influence of molecular size, the Zhdanov index has been normalized dividing by the number of atoms N. The information capacity of some classes of bioorganic compounds is thus calculated in bits per atom: carbohydrates, 0.25 to 0.33; aminoacids, 0.15 to 0.29; nucleotides, ≈ 0.2; B-group vitamins, ≈ 0.15; steroids, alkaloids, and terpenoids, < 0.1. Though the information contained in nucleotides is rather small it can be shown that the information capacity of nucleic acids is very high.

The active information capacity has been intensively used by Zhdanov for quantitative estimates of the specificity of organic compounds in various chemical reactions.

CHAPTER 6
Applications of the Molecular Information Theoretic Indices

1. ISOMER DISCRIMINATION

The problem of the isomer discrimination by means of a convenient numerical index is of importance for the structure-activity correlations, as well as for the coding and the computer processing of chemical structures. For the computer storage and retrieval of molecular systems, an index is needed which must characterize uniquely each of the isomeric molecules.

The detailed analysis of theoretic information indices, as well as of some topological indices (Bonchev et al. 1981 b) had shown that no index can discriminate isomeric molecules uniquely. A large sample of 427 graphs has been studied, encompassing all acyclic, monocyclic and bicyclic graphs having 4 to 8 vertices. The comparison between the different indices has been made by means of the mean isomer degeneracy i showing the mean number of isomers associated with one and the same value of the index:

$$i = N_{ind}/ N_{isom} \qquad (192)$$

Here N_{ind} is the number of distinct indices specified for the entire ensemble of N_{isom} isomers.

Table 15 shows that for acyclic graphs, even in the case of isomers with 8 carbon atoms, there are four information indices (the two indices on the graph distances, $^v\bar{I}_D^E$ and $^v\bar{I}_D^M$, and the information analogues of the Randić and Hosoya topological indices, $\bar{I}_{edge}^E$ and $\bar{I}_Z$, respectively) which discriminate uniquely the isomers. For cyclic graphs, however, degeneracy of the indices occurs even for. small graphs with 5 vertices. In the case of cyclic graphs with 8 vertices the best discriminating index $\bar{I}_Z$ has mean degeneracy i = 2.2.

TABLE 15. Mean degeneracy of 10 topological or topological information indices as calculated for 427 acyclic,
monocyclic, and bicyclic graphs having 4-8 vertices

Number of vertices in the isomer set	W	$^v\bar{I}_D^E$	$^v\bar{I}_D^M$	Z	$\bar{I}_Z$	x_R	$\bar{I}_{EDGE}^E$	$^v\bar{I}_C^E$	$^v\bar{I}_{CHR}^E$	$^v\bar{I}_{ORB}^E$
Acyclic Graphs										
4	1	1	1	1	1	1	1	1	1	1
5	1	1	1	1	1	1	1	1	1.5	
6	1	1	1	1	1	1	1	1	2	1.2
7	1.22	1	1	1	1	1	1	1.8	3.7	1.2
8	1.15	1	1	1.28	1	1.09	1	2.3	5.8	1.5
Cyclic Graphs										
4	1.5	1.5	1.5	1	1	1	1	1	1.5	1
5	2.5	2.0	2.0	1.7	1.4	1	1.4	2.0	3.3	2
6	3.6	2.4	2.2	2.6	1.5	1.2	1.6	4.1	5.8	3.6
7	6.0	2.8	2.4	4.5	2.2	1.6	3.4	12.1	12.1	9.4
8	10.6	3.5	2.8	7.7	2.2	1.6	6.0	16.0	36.4	18.2

[a]See Chapter IV for the index definitions

The most elaborate topological index, the so-called
distance connectivity (Chapter IV, Subsection 3.A.VI),
proposed quite recently by Balaban (1982) is considerably
more discriminating than the ten indices examined in
TABLE 15. Two degenerate pairs of bicyclic isomers appear
at N = 8, only. Yet, the validity of the conclusion about
the impossibility of a single structural index to discriminate uniquely isomers is unchanged.

The concept of a combined topological information index,
named superindex, developed in Chapter V, Section 4 (Bonchev et al., 1981 b), is a promising way towards the complete isomer discrimination. Eq. (189), representing a
simple version of the superindex, is sufficient to distinguish between all 427 isomers dealt with in TABLE 15.
Yet, bearing in mind the extremely fast increase in the
number of isomeric molecules with increasing number of
carbon atoms (e.g. the number of isomeric C_{25}-alkanes is
36797588!), more elaborate versions of the superindex
have been proposed (eq.188 and an equation its derivative).
They are far from solving the graph isomorphism problem,
indeed, but are sufficient for solving a very large number of isomer discrimination problems of practical importance (e.g. alkanes at least up to C_{30}). An advantage of
these more elaborate versions is that they are open systems admitting further extension when more complex molecules have to be discriminated.

2. CLASSIFICATION PROBLEMS

A. AROMATICITY OF MOLECULES

In Chapter III the information approach was shown to find
many applications in the classification of atoms and their

nuclei. Thus, information equations were shown to describe quantitatively the rows and groups in the Periodic Table of chemical elements. A new systematization of nuclides was proposed on information theoretic basis, and again information equations were demonstrated for the four known groupings of nuclides: isotopes, isotones, isobars, and isodifferent, as well as for a new, fifth such grouping called "isodefectants".

In the case of molecules, information equations were also presented for different homologous series of compounds on the basis of information indices of atomic composition, symmetry, nodal properties of the π-molecular orbitals, etc. In all these cases Information Theory is used mainly to express in a quantitative form the regularity existing within groups of objects that have been classified in advance. The information approach could, however, be of great help when classifying chemical systems, offering a common quantitative scale for their convenient comparison and grouping.

This possibility has been examined for the classification of molecules according to their aromatic properties (Fratev et al., 1980, 1982). Different definitions, as well as respective quantitative scales have been proposed for aromaticity. They divide the chemical compounds into three classes: aromatic, non-aromatic, and anti-aromatic compounds. The different aromaticity scales in general provide a similar ordering of chemical compounds though some discrepancies appear. They deal with the ground state of molecules, and to a very limited extent with the triplet excited state.

The information aromaticity indices I and $\tilde{I}$, defined in Chapter IV, Subsection 8. B by eq (173, 174), enable the correct qualifying of conjugated compounds as aromatic, non-aromatic, and anti-aromatic:

aromatic: $\quad \tilde{I} \approx 1 - 0{,}45, \ I \approx 0 - 0{,}06$

non-aromatic: $\quad \tilde{I} \approx 0.45 - 0.30, \ I \approx 0.06 - 0.095$

antiaromatic: $\quad \tilde{I} \approx 0.30 - 0, \quad I > 0.095$

The first class is characterized by low information content I, or by uniformity in the electron density distribution over atoms and bonds. This uniformity decreases, a bond alternation appears and gains in force when going from aromatic to non-aromatic, to antiaromatic molecules, in parallel with the increase in their information content. On the other hand, the information index $\tilde{I}$ of the non-aromatic non-alternant hydrocarbons is smaller than that of the corresponding alternant hydrocarbons, due to the uniform distribution of the π-electron charges in the latter.

It should be noted that the classification, made by Fratev et al. (1980) for a sample of 40 test compounds, coincides in principle with their classification made on the basis of other criteria of quite a different nature like resonance energy, topological resonance energy, conjugated π-electron circuits, etc. Differing from these approaches, however, the information aromaticity index is not deduced from any specific chemical or physical notions or models but automatically follows from the application of the Shannon equation to the entries of the charge - bond order matrix.

Similarly, the information treatment of ring aromaticity distinguishes three areas of

high : $\tilde{I} = 1 - 0.666,$

moderate: $\tilde{I} = 0.666 - 0.333$, and

low : $\tilde{I} = 0.333 - 0$, ring aromaticity.

The location of the examined cases within the above scale correlates numerically well with a number of other indices of ring aromaticity, defined within the molecular orbitals and valence bond methods, as well as with the Clar ideas for "empty" and "full" benzenoid fragments.

The information approach discussed above is the first unified approach to both total and ring aromaticity in the ground and excited states. It provided the first quantitative estimates of aromaticity in singlet excited states. A tendency towards similarity in the aromatic character of the aromatic and antiaromatic compounds has been established, resulting from the excitation from the ground to S_1 (fluorescent) state. Surprisingly, the benzene and cyclobutadiene molecules, located in the ground state at the two ends of the $\tilde{I}$ - scale, have been found to have in S_1 - state the same aromaticity index $\tilde{I}(1) = 0.355$.

Applying the information approach within the Hueckel molecular orbital approximation, Fratev et al. (1982) derived analytical expressions for the aromaticity index of 4n- and (4n + 2) - annulenes with and without bound alternation. A high correlation has been found between the information index and other aromaticity indices based on resonance energy, as well as with rate constants of the annulene formation reactions.

B. CLASSIFICATION OF BIOORGANIC COMPOUNDS

A new approach to the systematization of bioorganic compounds has been recently proposed by Zhdanov (1979). It is based on two different quantitative criteria. The active information capacity I_A of molecules, discussed in Chapter V, Section 5, is taken as their generalized structural chacteristic. It is essentially a dynamic molecular characteristic related to the ability of molecules to interact.

The second criterion is the *mean oxidation number* ω of the carbon atom in a bioorganic molecule. It is defined by averaging an algebraic sum of bond contributions: +1 and -1 for the carbon bond with more electronegative and more electropositive atoms, respectively:

$$\omega = \sum_i (C-X)_i / N_C \tag{193}$$

where N_c is the number of carbon atoms, and the summation is over all bonds formed by the carbon atoms. The crucial role that oxidation-reduction processes have in living things is regarded as sufficient reason for the choice of this parameter. In addition, the mean oxidation number has been found to correlate with a number of essential properties of bioorganic molecules, including here the genetic code (Zhdanov, 1974).

The generalized results of the Zhdanov classification are presented in FIG. 28.

Several points have to be noted here. The most important metabolites: carbohydrates, aminoacids, and vitamins are located close to the axis $\omega = 0$ (a metabolic axis). These compounds are not viewed suitable for an information transfer on a molecular level. This transfer is inherent

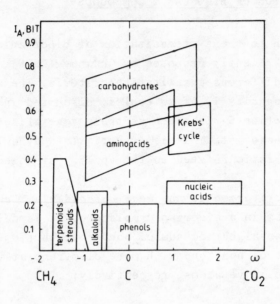

FIG.28. Diagram of bioorganic compounds (Zhdanov, 1979)

in nucleic bases and steroids, as well as in terpenoids and other compounds playing the role of external chemical information agents (attractants, odorants, etc.). All these compounds have a low active information capacity, due to their larger structural specificity.

The central place in the Zhdanov systematics is taken by aminoacids which combine in the same molecule acid and basic functions. To the left and right of aminoacids these two functions tend to be divided when going to the most reduced and most oxidized molecular forms, CH_4 and CO_2, respectively. Both bounds of the diagram are also occupied by some other products of organic compound decomposition, like uric acid, hydrocarbons, inorganic substances like NH_3, SH_2, etc.

The new classification has been used to correct the positioning of some bioorganic compounds within certain

classes of compounds, as, well as to discuss the molecular
aspects of evolution. The ageing process was found to be
associated with the metabolism displacement to the left of
the metabolic axis. It is likely that this combined information approach to bioorganic compounds reflects some
of their essential features, as well as evolution trends.

C. SYNTHESIS STRATEGY

Two similar approaches (Chapter V, Subsections 3. A and
3. B) to the quantitative estimate of the general molecular complexity have been used. A possibility is thus
provided for calculating the change in molecular complexity ΔC when going from simple reactants to a complex
target molecule. This can help in recognizing important
synthetic routes in the molecular design.

Bertz (1981) reported two such examples: the Diels-Al-
der reaction between butadiene and p-benzoquinone:

and the Weiss reaction between glyoxal and dimethyl 3-
ketoglutarate:

where the complexity of reaction species C_i is calculated according to eq (185).

The approach of Dosmorov (1980, 1982) is known in more detail. The molecular complexity (or otherwise, molecular information content) is calculated by eq. (186) for each of the inital, intermediate, and final reaction products, as well as for their sum I. The union of the latter forms the so-called family of isomeric ensembles of molecules (FIEM) where each isomeric ensemble corresponds to a certain state of the reaction system.

The set of all possible reaction mechanisms under study is generated proceeding from the assumption that they are informationally allowed, subject to the condition:

$$\frac{\Delta I}{\Delta <Q>} \geqslant 0 \qquad (194)$$

The above assumption is regarded plausible for reactions having low activation energy only.

As an illustration of the Dosmorov approach, the mechanisms generated for the chlorination of ethylene are shown:

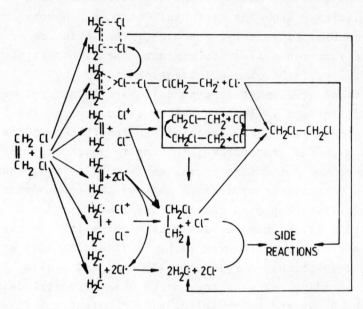

FIG.29. The set of informationally allowed mechanisms
   of the reaction $CH_2 = CH_2 + Cl_2 \rightarrow CH_2Cl\text{-}CH_2Cl$

3. SPECIFICITY OF CHEMICAL REACTIONS

Information indices can be applied to the characterization
of chemical reactions, as well. The analysis made by
Karreman (1955) on the information balance of chemical
reactions has already been discussed in Chapter IV,
Subsection 3. B. i on the basis of topological orbital
information. Further contribution has been made by Mowshovitz (1968 a) who derived a semiadditivity condition
for the orbital information which is of importance for
the information balance of chemical reactions.

   Zhdanov(1979) made use of information indices to characterize the specificity of chemical reactions. Let a number of reaction products be obtained in competing paral-

lel reactions. The Shannon function will indicate a lack
of specificity when the different reaction products are
equiprobable. Conversely, the higher the deviation of the
Shannon function H from its maximum, the higher is the
specificity of the chemical reaction.

Reactions of asymmetric synthesis have been first considered. In such a reaction, where two diastereomers A
and B are obtained, the ratio of the reaction products
varies with the temperature, tending to reach equilizati-
on. Thus, the H function increases (specificity decreases)
showing a behaviour similar to that of the thermodynamic
entropy. It is supposed that the information index of
stereoselectivity may be viewed as a chirality constant.

The above approach proved to be fruitful in cases of
aromatic substitution reactions (Zhdanov and Minkin,
1966) providing better agreement with experimental data
than the Brown and Smith selectivity constant S_f. Thus,
for the bromination of toluene where o-, m-, and p- isomers
are 33.3%, 0.3%, and 66.6%, respectively, the selectivity
constant $S_f = 2.664$. For the acetylation of toluene the
percentage is 1.1, 1.3, and 97.6, respectively, hence
$S_f = 2.192$. It follows from the comparison of the two
cases that toluene bromination is a reaction more specific
than acetylation. Clearly, this is not the case since
getting 97.6% of one isomer in the second reaction makes
it much more selective. The latter is adequately reflected
by the H-function which equals 0.2840 and 0.0564 for the
bromination and acetylation, respectively.

A number of other substitution reactions has been
studied such as toluene alkylation, nitration of various
monosubstituted benzenes, etc. Linear regression equation has been found between the information index H and
the Taft polar constant σ^* for nitration reactions, the
benzene substituents being NMe_3, NO_2, CH, F, Cl, COOH, Br,
I, COC_2H_5, CH_3CO, HO, COH_3, etc.:

$$H = 0.4937 - 0.09345 \; \sigma^{x}, \; r = 0.96 \qquad (195)$$

Similarly, for the chlorination of monosubstituted benzenes with substituents NO_2, CN, Cl, and Br:

$$H = 0.5923 - 0.0908 \; \sigma^{x}, \; r = 0.992 \qquad (196)$$

These linear regression equations provide evidence in favour of the supposed unified mechanisms of nitration and chlorination of the monosubstituted benzenes. Other conclusions have also been reached. Reactions are more specific in the case of a monoelectronic effect, as it is for the CH_3 and CCl_3 groups. When several effects are superimposed, the H index increases indicating the less specific (or more chaotic) reaction. The steric factor is also reflected by the H index which decreases with increasing substituent volume, predicting correctly the more specific reaction course. A general conclusion is made that the information approach could in many cases provide a new, deeper insight into the mechanisms of chemical reactions.

4. CORRELATIONS BETWEEN STRUCTURE AND PROPERTIES OF CHEMICAL COMPOUNDS

A. RULES FOR MOLECULAR STRUCTURES

The information theoretic indices can be applied to the characterization of the major structural features of molecules, their branching and cyclicity. One of the most promising approaches to molecular branching and cyclicity is the formulation of rules expressing the influence of the different structural factors. Dealing with branching factors, the branch number, length and location, the chain length, the number of branches linked with a certain chain vertex, etc., are considered. In the case of molecular cyclicity these are the cycle number, size and position, the degree of cycle condensation, the bending of a string of cycles, etc. The structure rules have been proved by inequalities expressing the regular change in the respective structural index during a characteristic molecular rearrangement. Besides their importance for chemical theory, the rules expressing the essence of molecular branching and cyclicity provide for the ordering of chemical structures, and primarily of isomeric molecules. This ordering is obeyed in principle by many physicochemical properties of chemical compounds. Thus, by using such structural rules, the chemist is in a position to predict approximately the values of various molecular properties without making any explicit calculations.

This approach is illustrated below by two rules of branching proved by means of the information index on the equality of the distances in the molecular graph, $^{v}\bar{I}_{D}^{E}$, defined in Chapter IV, subsection 3.F (Bonchev and Trinajstić, 1977). This index has been shown to be a convenient inversely proportional measure of molecular branching.

Rule 1. A tree is always more branched than a chain and less branched than a star having the same number of vertices.

| chain | tree | star |

Proving Rule 1 the following inequalities have been deduced:

$$^V\bar{I}_D^E \text{ (chain)} - {}^V\bar{I}_D^E \text{ (star)} = \sum_{k=1}^{N-2} k \; lb \; \sum_{k=1}^{N-2} k$$

$$- \sum_{k=1}^{N-2} k \; lb \; k > 0 \qquad (197)$$

where k stands for the different magnitudes of the distances occurring in the molecular graph.

$$^V\bar{I}_D^E \text{ (chain)} - {}^V\bar{I}_D^E \text{ (tree)} = (N-1) \; lb \; (N-1) -$$

$$- j \; lb \; j - (N-j-1) \; lb \; (N-j-1) > 0 \qquad (198)$$

where N is the total number of atoms in the molecule and j denotes the branch position along the main chain of atoms.

Rule 2. The relative branching increases when the increase in the total number of vertices of the graph increases the number of branches attached to a given vertex.

Example:

$$^V\bar{I}_D^E = 1 \; > \; 0.9710 \; > \; 0.9183 \; > \; 0.8631$$

It has been found as a consequence of Rule 1 that the normal alkanes out of all isomeric alkanes have maximum values of boiling point, heat of vaporization, heat of

combustion, and the vapour pressure coefficient from the
Antoine equation, and a minimum value of the heat of
formation of gaseous alkanes:

$$X_{normal\ alkanes} > X_{branched\ alkanes} \quad (X \equiv H^{298}_{comb,liq},$$

$$T_B, \lambda^o, b)$$

$$X_{normal\ alkanes} < X_{branched\ alkanes} \quad (X \equiv H^{298}_{el,gas}) \quad (199)$$

Such inequalities have been derived for numerous classes of acyclic and cyclic molecules as a consequence of
the derived rules of molecular branching and cyclicity
(Bonchev et al., 1978, 1980 a, Mekenyan et al., 1979 a,
1981 a,b, 1982). It should be noted however that most of
the rules of molecular branching and cyclicity have been
proven by means of the simpler Wiener index (Chapter IV,
Subsection 3.A.v) reflecting the structural factors quite
similarly to the information index of distances.

B. STRUCTURE-PROPERTY CORRELATIONS

Valentinuzzi and Valentinuzzi (1962, 1963) first reported
correlations between information indices and thermodynamic
molecular properties. They found the Morovitz index
(Chapter IV, Section 2) to correlate well with entropy
and heats of formation of various organic compounds. The
mean heat of formation, needed to generate 1 bit of information in the chemical structure, has been estimated
on this basis.

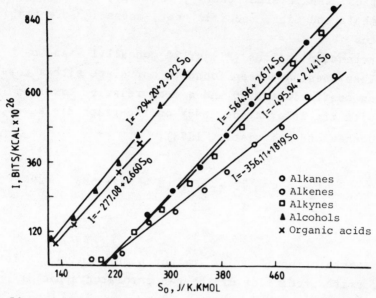

FIG.30. Variation of the information on molecular
symmetry with the absolute entropy for some homologous series of organic compounds

Similarly, the information on molecular symmetry (Chapter IV, Section 4) has been demonstrated to produce linear regression equations (FIG. 30) with the absolute entropies of alkanes, alkenes, alkynes, alcohols, and organic acids (Bonchev et al., 1976 c). The low relative error
(1-3%) had allowed the conclusion that the information:
entropy ratio is constant in a given homologous series,
which implies that every homologous series is characterized
by a constant degree of atomic organization.

A similar study on normal alkanes has shown (Bonchev and
Mekenyan, 1973), a linear regression equation between
entropy and a number of information indices, such as information on symmetry, atomic composition, and combined
indices of molecular complexity including individual atomic contributions (Chapter V, Section2). Linear regression lines have also been found with other thermodynamic

quantities: the internal energy, enthalpy, free energies of Helmholz and Gibs, specific heat, magnetic susceptibility, etc.

The retention indices (RI) of 28 monoalkyl - and o-dialkylbenzenes have been found to correlate with a correlation coefficient 0.99 and a mean relative error of 0.58% with the information index on the graph distances, $^{V}\overline{I}_D^E$ (Bonchev and Trinajstić, 1978):

$$RI = 2.97 \ ^{V}\overline{I}_D^E + 2.71 \ (N - 6) + 683 \qquad (200)$$

This correlation (FIG.31), rather simple in use, allows the accurate prediction of the gas chromatographic indices of alkylbenzenes.

The same information index has been used by Papazova et al. (1980), in combination with other structural parameters, for a correlation with the gas chromatographic retention times of 56 isoalkanes:

$$RI = 128.8 - 12.9 \ ^{V}\overline{I}_D^E - 21.6 \ n_R + 21.4 \ n_B + 57.8 \ n_o -$$

$$- 12.5 \ n_i + 16.8 \ n_L, \ r = 0.9986 \qquad (201)$$

Here n_R, n_B, n_o, n_i, and n_L stand for the number of substituents, number of butane chains, total number of carbon atoms, number of carbon atoms in the substituents, and number of carbon atoms in the straight chain of the alkanes, respectively. The mean deviation of RI_{calc} from RI_{exp} is less than 4 i.u. which provides prediction of the retention indices of higher homologs with a satisfactory accuracy. The significance of the chosen structural

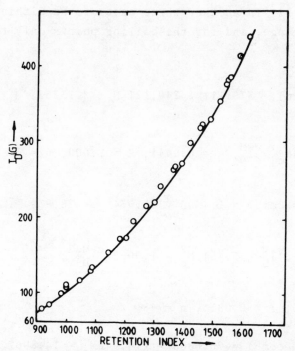

FIG.31. Correlation between the information on the
graph distances and the chromatic retention index
of 28 mono- and o- dialkylbenzenes

elements has been proven by extending the initial number
of 56 isoalkanes to 118.

Extensive correlations have been reported by Mekenyan
et al. (1980) between the information on the magnitude of
the graph distances $^{V}I_D^M$ and a number of properties of alkylbenzenes: heats of formation, ΔH^o_{298} (form); heats of
combustion in liquid and gas phase, ΔH^o_{298} (gas, comb)
and ΔH^o_{298} (liq, comb); heats of vaporization, ΔL_{298};
molecular volume, V^{20}; boiling point, T^o_B; and parachor, P.
The total number of carbon atoms N_C has been used in the
correlations as a second variable. The correlation coefficient r has been found to be equal or close to 1. In three
cases the mean relative error ε has been determined to be

less than 0.1%, in another three cases to be within
0.2 - 0.4% range, and for the boiling points only to be
1.25%:

$$\Delta H^o_{298} \text{ (form)} = 375.111 - 240.121 \, N_C + 2.13262 \, {}^V\bar{I}^M_D -$$

$$- \, 0.025364 N_C \cdot {}^V\bar{I}^M_D, \quad \varepsilon = 0.044\%, \quad R = 1.000, \quad n = 48 \qquad (202)$$

$$\Delta H^o_{298} \text{ (gas comb)} = 1.4309 - 85.034 N_C - 28.934 \, {}^V\bar{I}^M_D$$

$$+ \, 5.5986 N_C \cdot {}^V\bar{I}^M_D - 7.4543 N^2_C - 1.0002 ({}^V\bar{I}^M_D)^2;$$

$$R = 1.0000, \quad \varepsilon = 0.081\%, \quad n = 48 \qquad (203)$$

$$\Delta H^o_{298} \text{ (liq comb)} = 70.6868 - 118.946 N_C - 14.8967 \, {}^V\bar{I}^M_D$$

$$- \, 1.5674 N^2_C + 0.4291 \, ({}^V\bar{I}^M_D)^2;$$
$$R = 1.0000, \quad \varepsilon = 0.070\%, \quad n = 36 \qquad (204)$$

$$V^{20} = - \, 189.3186 + 129.74 N_C - 52.320 \, {}^V\bar{I}^M_D + 15.277 N_C \, {}^V\bar{I}^M_D -$$

$$- \, 17.021 N^2_C - 3.3586 \, ({}^V\bar{I}^M_D)^2; \quad R = 0.9991, \quad \varepsilon = 0.41\%,$$

$$n = 21 \qquad (205)$$

$$T^o_B = - \, 60.6342 - 4.9950 N_C \, {}^V\bar{I}^M_D + 9.7833 N^2_C + 0.072 \, ({}^V\bar{I}^M_D)^3$$

$$- \, 0.02295 N^4_C; \quad R = 0.9947, \quad \varepsilon = 1.25\%, \quad n = 29 \qquad (206)$$

$$\Delta L_{298} = 0.2150 + 1.7502N_c - 0.3931 \ {}^V\bar{I}_D^M; \ R = 0.9978,$$

$$\varepsilon = 0.42\%, \ n = 13 \tag{207}$$

$$P = -19.5077 + 31.075N_c + 6.7209 \ {}^V\bar{I}_D^M - 0.1358 \ ({}^V\bar{I}_D^M)^2,$$

$$R = 0.9998, \ \varepsilon = 0.22\%, \ n = 21 \tag{208}$$

where n is number of compounds in the correlation sample.
The correlation with the heats of formation has been additionally amended by dividing the compounds into three groups: mono-, di-, and trialkylbenzenes:
Monoalkylbenzenes:

$$\Delta H_{298}^o \ (form) = +374.36 + 1.827 \ {}^V\bar{I}_D^M - 239.83N_c -$$

$$- 0.0279 \ ({}^V\bar{I}_D^M)^2 + 0.0059N_c^3, \ R = 1.0000, \ \varepsilon = 0.012\%,$$

$$n = 20 \tag{209}$$

Dialkylbenzenes:

$$\Delta H_{298}^o \ (form) = +325.71 - 2.858 \ {}^V\bar{I}_D^M - 227.16N_c +$$

$$+ 0.0025 \ ({}^V\bar{I}_D^M)^3 - 0.0185N_c^3$$

$$R = 1.0000, \ \varepsilon = 0.009\%, \ n = 12 \tag{210}$$

Trialkylbenzenes:

$$\Delta H^{o}_{298} \ (\text{form}) = + \ 469.54 - 26.076 \ ^{V}\overline{I}^{M}_{D} - 228.02N_{c} +$$

$$+ \ 0.8906 \ (^{V}\overline{I}^{M}_{D})^{2}, \ R = 1.0000, \ \epsilon = 0.011\%,$$

$$n = 9 \tag{211}$$

As seen, the mean relative error is thus strongly reduced, as compared with the initial equation (202). It is
slightly less than that in the Tatevskii (1953) additive
scheme ($\epsilon=0.013\%$) where the same sample of compounds is
partitioned into 11 subgroups (Tatevskii, 1953). For each
such subgroup of alkylbenzenes the Tatevski scheme usually requires 15 empirical parameters while in the above
correlations they are usually four. Correlations such as
eq (202) to (211) indicate that the molecular indices
could predict the thermodynamic properties of chemical
compounds with an accuracy comparable or higher than that
of the well known Tatevskii scheme.

The electropy and bondtropy indices introduced in
Chapter IV, Subsection 8.A, have also been a subject of
various correlations with thermodynamic properties of
hydrocarbons having up to 10 carbon atoms (Sakamoto et
al., 1977). Linear relationships have been found between
these indices and the extensive intramolecular quantities
standard entropy, S^{o}, heat of formation ΔH^{o}_{f} and combustion ΔH^{o}_{c}, and standard Helmholz energy of formation ΔF^{o}_{f}
of normal alkanes and 1-alkenes, as well as with the ex-
ception of S_{o} and ΔH^{o}_{f} - for cycloalkanes. Two examples
dealing with normal cycloalkanes are given below:

$$- \Delta H_f^o \text{ (gas)} = 4.4893\varepsilon + 247.5359, \ r = 0.9996 \tag{212}$$

$$- \Delta H_c^o \text{ (liq)} = 4.3878\varepsilon + 258.5097, \ r = 0.9998 \tag{213}$$

A non-linear correlation has been obtained between electropy and boiling point T_B which is an intensive intermolecular quantity:

Normal alkanes:

$$\varepsilon = 0.0020 \ T_B^2 + 0.8091 \ T_B + 85.6720 \tag{214}$$

where the multiple correlation coefficient $r = 0.9999$, and the partial correlation coefficients are $r_1 = 0.9974$ for T_B^2 and $r_2 = 0.9998$ for T_B, respectively.

Normal cycloalkanes:

$$T_B = (-6.5003 + (0.6948\varepsilon - 8.2036)^{1/2})/0.0347 \tag{215}$$

where $r = 0.9995$, $r_1 = 0.9664$, and $r^2 = 0.9913$

Normal 1-alkenes:

$$T_B = (-7.7231 + (0.8852\varepsilon - 17.8543)^{1/2})/0.0443 \tag{216}$$

where $r = 1.0000$, $r_1 = 0.9990$, $r_2 = 0.9999$

The last three correlations with normal structures are called main correlations. In addition, linear subcorrelations of $T_B = a\varepsilon + b$ type have been obtained with high correlation coefficients for the isomeric group of hydr-

ocarbons (FIG.32). These linear subcorrelations are of
the same slope (a ≈ const), while the parameter b depends on the number of carbon atoms.

Kier (1980) examined for a number of alcohols the correlation between the heat of vaporization ΔL_{298} and the
orbital information index I (Chapter IV, Subsection 3.B.i),
called by him "molecular negentropy":

$$\Delta L_{298} = 0.265 \ (\pm 0.000)I + 8.566 \ (\pm \ 0.29) \tag{217}$$

$$n = 15, \ r = 0.989, \ s = 0.414, \ F = 563.$$

Here, as well as in Subsection C, n is the number of da-

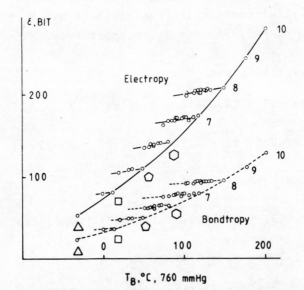

FIG.32. Correlation between the electropy or bondtropy ε and boiling point T_B of cycloalkanes
C_nH_{2n} (Sakamoto et al., 1977)

ta points, r the correlation coefficient, s the standard
deviation, and F the F- ratio between the variance of the

ob'served and calculated values.

The high correlation obtained is viewed as an evidence supporting the supposed connection between molecular information content (molecular negentropy) and dispersion forces.

C. STRUCTURE-ACTIVITY CORRELATIONS

Contemporary drug design is largely based on the connection between the structure of a molecule and its biological and pharmacological activity. Various approaches have been used in the search for quantitative structure-activity relationships (QSAR), (See Purcell et. al., 1973; Martin, 1978). Recently, the possibility of expressing the molecular structure by means of a numerical topological index has attracted much attention (Kier and Hall, 1976; Balaban et al., 1980; Sablić and Trinajstić, 1981). The topological indices (Chapter IV, Subsection 3.A) reflect the size, shape, branching and cyclicity patterns, as well as bonding type of a molecule.

The various topological and electronic information indices may find wide applications to the quantitative structure-activity relationships. Being flexible structural characteristics, capable of reflecting any details of the molecular geometric and electronic structure, these indices seem to be the most appropriate heuristic approach for such purposes.

In terms of Information Theory a biologically active molecule might be viewed as a message, composed of atoms and bonds (or atomic nuclei and electrons). The information contained in this message transmits to a corresponding receptor, causing a biological response.

Kier (1980) applied the orbital information index and

that on molecular symmetry (Chapter IV, Subsection 3.B.i,
and Section 4, respectively) to correlations with enzyme
inhibition, nonspecific narcotic activity, and toxicity
of various alcohols. Thus for nonspecific narcotic agents
(10 alcohols, 4 ketones, 2 esters, 2 carbamates, one
nitrile and one oxime) the orbital information index I
has been found to correlate well with the narcotic potency
(log 1/C):

$$\log 1/C = 0.102 \ (\pm 0.000)I + 0.130 \ (\pm 0.09) \qquad (218)$$

$$n = 32, \ r = 0.975, \ s = 0.187, \ F = 582$$

An example for the relationship between the information
index and *in vitro* potencies pK_I of the enzyme lipoxyg-
enase is also given:

$$pK_I = 0.108 \ (\pm 0.000) \ I - 0.450 \ (\pm 0.12) \qquad (219)$$

$$n = 12, \ r = 0.986, \ s = 0.140, \ F = 353$$

Extensive studies have been carried- out by an Indian
group of researchers (Basak, Ghosh, Ray, Raychaudhury,
and Roy) who apply the information indices IC, SIC, BIC,
etc. (Chapter IV, Subsection 3.J.ii) developed for multigraphs. Thus, Raychaudhury et al. (1980) made use of IC
and SIC indices in correlations with the biological action of spasmolytics, as well as barbiturates and they obtained results with a high degree of statistical significance. The same indices have been correlated (Ray et al.,
1981a) with the antipsychotic action of sintamil, diazepam, chloropromazine, etc., with the pharmaclological parameters of diphenhydramine series, as well as with the

pharmacological properties of primary aliphatic alcohols. Some of their results are given below.

Local anesthetic activity (LAA) of aliphatic alcohols:

$$LAA = -3.1259 + 9.6143 \ (SIC) \tag{220}$$

$$(n = 8, \ r = 0.9732, \ s = 0.3028, \ F_{1,6} = 107.3324);$$

Narcotic activity of barnacle larvae (PC_2) of aliphatic alcohols:

$$PC_2 = 6.1657 - 9.5720 \ (SIC) \tag{221}$$

$$(n = 8, \ r = 0.9826, \ s = 0.2417, \ F_{1,6} = 167.4664);$$

Antihistaminic activity (AHA) of diphenhydramines:

$$AHA = 16.1979 - 7.2952 \ (IC) \tag{222}$$

$$(n = 5, \ r = 0.9900, \ s = 0.0662, \ F_{1,3} = 148.2789)$$

Antipsychotic activity (ED_{50}) of sintamil, imipramine, reserpine, diazepam, chloropromazine, benzoctamine, and chlorodiazepoxide:

$$ED_{50} = 2269.357 - 9263.6818(SIC) - 9425.9932(SIC)^2 \tag{223}$$

$$(n = 7, \ r = 0.8488, \ s = 14.6295, \ F_{2,4} = 5.1549).$$

The BIC index has been applied to QSAR of barbiturates and alkyl carbamates (Ray et al., 1981 b), as well as of

experimental odour intensity of a group of fatty acids
(Ray et al., 1981 c). The last study showed the effectiv-
eness of this index in predicting the odour quality of
molecules.

The diazepam antagonizing potency (I_{50}) and the convul-
sive dose (CD) of tetrazols have been successfully predicted by the IC index:

$$\text{Log } I_{50} = -27.6845 + 22.6835 \text{ (IC)} - 4.4408(\text{IC})^2 \qquad (224)$$

$$(r = 0.9932);$$

$$\text{Log CD} = 13.4280 - 16.5965(\text{IC}) + 4.7357(\text{IC})^2 \qquad (225)$$

$$(r = 0.9928).$$

Similar correlations have been reported (Ray, 1982 a)
between the SIC index and some physical and biological
properties of N-alkylnorketobemidones and triazinones.
Thus, for *in vivo* analgesic potencies (A-ED$_{50}$) and in
vitro receptor binding abilities (no-sodium ED$_{50}$ nmol/1)
of norketobemidones:

$$\text{A-ED}_{50} = 4016.5278 - 22513.001(\text{SIC}) + 31549.105(\text{SIC})^2$$
$$(226)$$
$$(n = 7, \ r = 0.9282, \ s = 10.4606, \ F_{2,4} = 12.4378);$$

$$\text{no-sodium ED}_{50} = 44092.812 - 244112.27(\text{SIC}) +$$

$$+ 337570.31(\text{SIC})^2 \qquad (227)$$

$$(n = 6, \ r = 0.9077, \ s = 120.7523, \ F_{2,4} = 11.7020)$$

The F-values in eq (224, 225) are significant at the
97.5% level.

The same indices have also been applied by Ray et al.
(1982 b) to three important series of bioactive agents:
triazinones, alkylcarbamates and β-pyridylalkylamines.

A newly formulated information theoretic index - complementary information content (CIC) - has been applied
to the modelling of thermodynamic properties of congeneric
alcohols (Raychaudhury et al., 1982), as well as (Basak
and Magnuson, 1982) to the study of their narcosis
activity (log LC_{50}) and lipophilicity (log P):

$$\log LC_{50} = -1.8964(CIC) + 1.9788 \qquad (228)$$

(n = 10, r = 0.9889, s = 0.3233, $F_{1,8}$ = 355.33, 99.9%
level of significance).

$$\log LC_{50} = - 0.9394 (\log P) - 0.8416 \qquad (229)$$

(n = 10, r = 0.9955, s = 0.2073; $F_{1,8}$ = 875.94).

Further work is in progress by the Calcutta group on
antifungal compounds like Cinnamic acid derivatives, γ -
pyron derivates, etc.

Concluding this Subsection, it should be noted that
though "potentially very usable" (Sablić and Trinajstić,
1981) the information indices have just started their
long way in drug design. The large number of indices, introduced in Chapter IV and V, are still in the explorat-
ion stage as a tool in predicting pharmacological activity of chemical compounds. Some limitations of the known
indices have also to be taken into account. Reflecting

the structure of molecule as a whole these indices are mainly appropriate for correlations with non-specific interactions. The specific biological activity, determined by some molecular fragments or functional groups, needs to be described by a new type of indices that are not global. This is a problem to be solved though some possible trends have already emerged. Such an example is the information index on electron delocalization (aromaticity) which can be calculated for separate rings in a polycyclic molecule (Chapter IV, Subsection 8.B). It may be expected that the fruitfulness of the information approach to QSAR studies would be demonstrated in the near future.

Bibliography

Alhassid, Y. and Levine, R.D. (1977). Entropy and chemical change. III. The maximal entropy (subject to con-
straints) procedure as a dynamical theory. J. Chem. Phys.
67, 4321 - 4339.

Andrews, D.H. and Boss, M.L. (1971). Transfer of information in growth processes. Yale Sci. *45*, 2 - 11.

Ashby, W. (1956). *An Introduction to Cybernetics*. Wiley,
New York.

Aslangul, C., Constanciel, R., Daudel, R., and Kottis,
P. (1972). Aspects of the localizability of electrons
in atoms and molecules: loge theory and related methods.
Adv. Quantum Chem. *6*, 94 - 141.

Balaban, A.T., Farcasiu, D., and Banica, R. (1966).
Graphs of multiple 1,2-shifts in carbonium ions and
related systems. Rev. Roum. Chim. *11*, 1205 - 1227.

Balaban, A.T. (1975). Some chemical applications of
Graph Theory. Math. Chem. (MATCH) *1*, 33 - 60.

Balaban, A.T., Editor (1976). *Chemical Applications of
Graph Theory*. Academic, New York.

Balaban, A.T. (1979). Chemical graphs. XXXIV. Five new
topological indices for the branching of tree-like
graphs. Theor. Chim. Acta *53*, 355 - 375.

Balaban, A.T. and Motoc, I. (1979). Chemical graphs.
XXXVI. Correlations between octane numbers and topological indices of alkanes. Math. Chem. (MATCH) *5*, 197 - 218.

Balaban, A.T., Chiriac, A., Motoc, I., and Simon, Z.
(1980). *Steric Fit in QSAR*. Lecture Notes in Chemistry,
No. 15, Springer, Berlin.

Bartel, H.G.(1975). Estimation of the direction of benzene
derivatives. Z. Chem. *15*, 62 - 63 (in German).

226

Balaban, A.T. Bonchev D., Mekenyan, Ov., and Motoc, I.
(1982), in Steric effects in structure-activity relations.
Top. Curr. Chem. (in press).

Balaban, A.T. (1982). Highly discriminating topological
index. J. Am. Chem. Soc. (submitted).

Basak, S.C., Roy, A.B., and Ghosh, J.J. (1979), in
Proceedings of the 2nd Intern. Conf. on Math. Modelling.
p. 851, Univ. of Missouri, Rolla.

Basak, S.C., Ray, S.K., Raychaudhury C., Roy, A.B., and
Ghosh, J.J. (1982). Molecular topology and pharmacological action: A QSAR study of tetrazoles using topological
information content (IC). IRCS Medical Science: Biochem-
istry; Drug Metabolism and Toxicology; Pharmacology, *10*,
145 - 146.

Basak, S.C. and Magnuson, V.R. (1982). Molecular topology
and narcosis: A quantitative structure - activity relationship (QSAR) study of alcohols using complementary
information content (CIC). Arzneim. Forsch. (in press).

Bawden, D., Catlow, J.T., Devon, T.K., Dalton, J.M.,
Lynch, M.F., and Willett, P. (1981). Evaluation and
implementation of topological codes for online compound
search and registration. J. Chem. Inf. Comput. Sci. *21*,
83 - 86.

Behzad, M. and Chartrand, G. (1972). *Introduction to the
Theory of Graphs.* Allyn and Bacon, Boston.

Ben - Shaul, A., Levine, R.D., and Bernstein, R.B. (1972).
Entropy and chemical change. II. Analysis of product
energy distributions: temperature and entropy deficiency.
J. Chem. Phys. *57*, 5427 - 5447.

Bernstein, R.B. and Levine, R.D. (1972). Entropy and
chemical change. I. Characterization of product (and
reactant) energy distributions in reactive molecular
collisions: Information and entropy deficiency. J. Chem.
Phys. *57*, 434 - 449.

Bertz, S.H. (1980). Abstracts, Rep. No. 501, Southeast Southwest Regional Meeting of the American Chemical Society, New Orleans.

Bertz, S.H. (1981). The first general index of molecular complexity. J.Am. Chem. Soc. *103*, 3599 - 3601.

Beskov, V.S. and Solyakina (1975). Sensitivity and information content of continuous and continuous-circulation methods for measuring the catalytic activity. Tr. N.-i. i Proekt. In-ta Azot. Prom-sti i Produktov Organ. Sinteza, No. 36, 49 - 59. (in Russian).

Boltzmann, L. (1896). *Vorlesungen über Gastheorie*. Leipzig.

Bonchev, D. (1970). Thermodynamics, information, and evolution of matter. Biologiya i Khimiya, Sofia *13*, 4 - 11 (in Bulgarian).

Bonchev, D. and Peev, T. (1973). A theoretic-information study of chemical elements. Mean information content of a chemical element. God. Vissh Khim. - Technol. Inst., Burgas, Bulg. *10*, 561 - 574 (in Bulgarian).

Bonchev, D., Tashkova, C., and Ljuzkanova, R. (1975). On the correlation between enthalpy of formation, atomic number, and information content of alkali halides. Dokladi BAN *28*, 225 - 228.

Bonchev, D., Peev, T., and Rousseva, B. (1976 a). Information study of atomic nuclei, Information for proton-neutron composition. Math. Chem. (MATCH) *2*, 123 - 137.

Bonchev, D., Kamenska, V., and Tashkova, C. (1976 b). Equations for the elements in the Periodic Table, based on Information Theory. Math. Chem. (MATCH) *2*, 117 - 122.

Bonchev, D., Kamenski, D., and Kamenska, V. (1976 c). Symmetry and information content of chemical structures. Bull. Math. Biol. *38*, 119 - 133.

Bonchev, D., Kamenska, V., and Kamenski, D. (1977). Information content of chemical elements. Monatsh. Chem. *108*, 477 - 487 (in German).

Bonchev, D. and Lickomannov, G. (1977 a). Information theory, thermodynamics, and statistics (Selected bibliography). Math. Chem. (MATCH) *3*, 269 - 273.

Bonchev, D. and Lickomannov, G. (1977 b). Statistical properties of finite point groups. Math. Chem. (MATCH) *3*, 3 - 20.

Bonchev, D. and Trinajstić, N. (1977). Information theory, distance matrix, and molecular branching. J. Chem. Phys. *67*, 4517 - 4533.

Bonchev, D. and Kamenska, V. (1978 a). Information theory in describing the electronic structures of atoms. Croat. Chem. Acta *51*, 19 - 27.

Bonchev, D. and Kamenska, V. (1978 b). Information characteristics of periods and subperiods in the Periodic Tabble. Monatsh. Chem. *109*, 551 - 556 (in German).

Bonchev, D., Mekenyan, O., Knop, J.V., and Trinajstić, N. (1978), Topological study of monocyclic structures. Croat. Chem. Acta *52*, 361 - 367.

Bonchev, D., and Trinajstić, N. (1978). On topological characterization of molecular branching. Int. J. Quantum Chem. Symp. *12*, 293 - 303.

Bonchev, D. (1979). Information indices for atoms and molecules. Math. Chem. (MATCH) *7*, 65 - 113.

Bonchev, D. and Kamenska, V. (1979 a). Electron Transitions and the change in the information content of chemical elements. Monatsh. Chem. *110*, 607 - 612 (in German).

Bonchev, D. and Kamenska, V. (1979 b). Correlations between information indices and properties of chemical elements. Math. Chem. (MATCH) *7*, 113 - 133.

Bonchev, D., Knop, J.V., Trinajstić, N. (1979 a). Mathematical models of branching. Math. Chem. (MATCH) *6*, 21 - 47

Bonchev, D. and Lickomannov, G. (1979). Information theory analysis of the nodal properties of Π-molecular orbitals. Acta Chim. Acad. Sci. Hung. *102*, 321 - 332.

Bonchev, D., Mekenyan, O., Protić, G., and Trinajstić, N. (1979 b). Application of topological indices to gas chromatographic data: Calculation of the retention indices of isomeric alkylbenzenes. J. Chromatogr. *176*, 149 - 156.

Bonchev, D., Mekenyan, O., and Trinajstić, N. (1980 a). Topological characterization of cyclic structures. Int. J. Quantum Chem. *17*, 845 - 893.

Bonchev, D., Balaban, A.T., and Mekenyan, O. (1980 b) Generalization of the graph centre concept, and derived topological indices. J. Chem. Inf. Comp. Sci. *20*, 106-113.

Bonchev, D. (1981). Information theory interpretation of the Pauli principle and Hund rule. Intern. J. Quant. Chem. *19*, 673 - 679.

Bonchev, D. and Kamenska, V. (1981). Predicting the properties of the 113 - 120 transactinide elements. J. Phys. Chem. *85*, 1177 - 1186.

Bonchev, D., Balaban, A..T., and Randić, M. (1981 a). The graph centre concept for polycyclic graphs. Int. J. Quantum Chem. *19*, 61 - 82.

Bonchev, D. Mekenyan, O., and Trinajstić, N. (1981 b). Isomer discrimination by topological information approach. J. Comp. Chem. *2*, 127 - 148.

Bonchev, D. and Gruncharov, I. (1982). The program "DISTANCE" for centroidal ordering and canonical numbering of graph vertices and edges. J. Chem. Inf. Comp. Sci. (to be submitted).

Bonchev, D. and Mekenyan, O. (1982 a). Theoretic information indices for graphs, multigraphs, and directed graphs. MATCH (To be submitted).

Bonchev, D. and Mekenyan, O. (1982 b). On the complexity of graphs used in chemical technology. Int. Chem. Eng. (to be submitted).

Bonchev, D. and Temkin, N. (1982). On the complexity of chemical reactions with a linear mechanism. J. Comp.

230

Chem. (To be submitted).

Brekhman, I.I. (1976). *Man and Biologically Active Substances*. Nauka, Moscow (in Russian).

Brillouin, L. (1956). *Science and Information Theory*. Academic, New York.

Brillouin, L. (1964). *Scientific Uncertainty and Information*. Academic, New York.

Bykhovskii, A.I. (1968). Negentropy principle of information and some problems in bioenergetics. Math. Biosci. *3*, 353 - 370.

Chart of the Nuclides, Sec. Edition (1963). Gersbach, Muenchen.

Cotton, F.A. (1971). *Chemical Application of Group Theory*. 2nd ed., Wiley Interscience, New York.

Dancoff, S.M. and Quastler, H. (1953), in *Essays on the Use of Information Theory in Biology*. (Edited by H. Quastler). University of Illinois, Urbana.

Dimov, D. and Bonchev, D. (1976). Spin-information equations of the groups and periods in the Periodic Table of chemical elements. Math. Chem. (MATCH) *2*, 111 - 115.

Dosmorov, S.V. (1982). Generation of homogeneous reaction mechanisms. Kinet. Katal., in press, (in Russian).

Dosmorov, S.V. (1980). in *Fifth All-Union Conference "Use of Computers in Molecular Spectroscopy and Chemical Research", Tezissi Dokladov* (Edited by M. I. Podgornaya), p. 28. SO AN SSSR, Novosibirsk (in Russian).

Eckschlager, K., and Štěpánec, V. (1979). *Information Theory as Applied to Chemical Analysis*. Wiley, New York.

Eigen, M. and Winkler, P. (1975). *Lidus Vitalis*. Boehringen, Mannheim.

Eigen, M. and Schuster, P. (1979). The Hypercycle. A Principle of Natural Self Organization. Springer, Berlin.

Evans, L.A., Lynch, M.F., and Willett, P. (1978). Structural search codes for online compound registration. J. Chem. Inf. Comput. Sci. *18*, 146 - 149.

Fischer, R.A. (1925). Theory of statistical estimation. Proc. Camb. Phil. Soc. *22*, 700 - 725.

Fleming, I. (1976). *Frontier Orbitals and Organic Chemical Reactions*. Wiley, London.

Fratev, F., Bonchev, D., and Enchev, V. (1980). A theoretic information approach to ring and total aromaticity in ground and excited state. Croat. Chem. Acta *53*, 545-554

Fratev, F., Polansky, O.E., Bonchev, D., and Enchev, V. (1982). A theoretic information study on the electron delocalization (aromaticity) of annulenes with and without bond alternation. Theochem (in press).

Freeland, R.G., Funk, S.A., O'Korn, L.J., and Wilson, G.A. (1979). The Chemical Abstract Service chemical registry system. II. Augmented connectivity molecular formula. J. Chem. Inf. Comput. Sci. *19*, 94 - 98.

Gamov, G. (1954 a). Possible relation between deoxyribonucleic acid and protein structure. Nature *173*, 318 - 319.

Gamov, G. (1954 b). Possible mathematical relation between deoxyribonucleic acid and proteins. Dann. Biol. Medd. *22*, 1 - 13.

Gatlin, L. (1972). *Information Theory and the Living System*. Columbia University Press.

Glushkov, V. (1964), in *Cybernetics, Thinking, and Life* (Edited by A.I. Berg). p. 53, Misl. Moscow (in Russian).

Gordon, M., and Scantlebury, G.R. (1964). Non-random polycondensation: Statistical theory of the substitution effect. Trans. Faraday Soc. *60*, 604 - 621.

Gordon, M. and Temple, W.B., (1970). Chemical combinatorics. Part I. Chemical kinetics, graph theory, and combinatorial entropy. J. Chem. Soc. A, 729 - 737.

Gutman, I. and Trinajstić, N. (1973). Graph theory and molecular orbitals. Top. Curr. Chem. *42*, 49 - 93.

Gutman, I., Ruŝĉiĉ, B., Trinajstić and Wilcox, C.F., Jr. (1975) Graph theory and molecular orbitals. XII. Acyclic polyenes. J. Chem. Phys. *62*, 3339 - 3405.

232

Harary, F. (1969). *Graph Theory*. Addison-Wesley, Reading, Mass.

Hardy, G.H., Littlewood, J.E., and Polya, G. (1934). *Inequalities*. Cambridge University, London.

Hasegawa, M. and Yano, T.A. (1975). Entropy of the genetic information and evolution. Origins Life *6*, 219 - 227.

Hochstrasser, R. (1976). *Molecular Aspects of Symmetry*. Benjamin New York.

Hosoya, H. (1971). A newly proposed quantity characterizing topological nature of structural isomers of saturated hydrocarbons. Bull. Chem. Soc. Jpn. *44*, 2332 - 2339.

Hosoya,H. (1972). Topological index as a sorting device for coding chemical structures. J. Chem. Doc. *12*, 181-183.

Ingarden, R.S. and Urbanik, K. (1961). Information as a fundamental notion of Statistical Physics. Bull. Acad. Sci. Polon., Ser. sci. math. *9*, 313 - 316.

Jaffé, H.H. and Orchin, M. (1965). *Symmetry in Chemistry*. Wiley, New York.

Jaynes, E.T. (1957 a). Information theory and statistical mechanics.
Phys.Rev. *106*, 620 - 630; idem. (1957 b). Information theory and statistical mechanics. II.
Phys.Rev. *108*. 171 - 190.

Karreman, G. (1955). Topological information content and chemical reactions. Bull. Math. Biophys. *17*, 279 - 285.

Kawasaki, K., Mizutani, K., and Hosoya, H. (1971). Tables of non-adjacent numbers, characteristic polynomials and topological indices. II. Mono and bicyclic graphs. Nat. Sci. Rept. Ochanomizu Univ. (Jpn) *22*, 181 - 214.

Khinchin, A.I. (1957). *Mathematical Foundations of Information Theory*. Dover, New York.

Kier, L.B., Hall, L.H., Murray, W.J., and Randić, M. (1975). Molecular connectivity I: Relationship to nonspecific local anesthesia. J. Pharm. Sci. *64*, 1971 - 1974.

Kier, L.B. and Hall, L.H. (1976). *Molecular Connectivity in Chemistry and Drug Research*. Academic, New York.

Klechkovski, V. (1968). *Distribution of Atomic Electrons and the Rule of Subsequent Filling of the (n+1) - Groups* Atomisdat, Moscow (in Russian).

Kobozev, N.I. (1971). *A Study on the Thermodynamics of Information and Thinking Processes*. Moscow University, Moscow (in Russian).

Kobozev, N.I., Strakhov, B.V., and Rubashev, A.M. (1971 a). Information theory applications to the study of catalytic systems. I. Crystal catalysts. Zh. Fiz. Khim. *45*, 86 - 89; idem. (1971 b). II. Crystals with defects. absorption catalysts. Zh. Fiz. Khim. *45*, 375 - 378 (in Russian).

Kolmogorov, A.N. (1965). Three approaches to the notion "quantity of information". Problemi peredachi informatsii *1*, 3 - 11 (in Russian).

Kolmogorov, A.N. (1969). On the logical foundations of information theory and probability theory. Problemi peredachi informatsii *5*, 3 - 7 (in Russian).

Kravtzov, V. (1974). *Atomic Masses and Nuclear Binding Energies*. Atomisdat, Moscow (in Russian).

Kullback, S. and Leibler, R.A. (1951). On information and sufficiency. Ann. Math. Stat. *22*, 79 - 86.

Larson, E.G. (1973). Representable reference density matrices induced by Information Theory. Int. J. Quantum Chem. *7*, 853 - 867.

Lederberg, J., Sutherland, G.L., Buchanan, B.G., Feigenbaum, E.A., Robertson, A.V., Duffield, A.M., and Djerassi, C. (1969). Application of artificial intelligence for chemical inference. I. The number of possible organic compounds. Acyclic structures containing C, H, O, and N.J. Am. Chem. Soc. *91*, 2973 - 2976.

Levine, R.D. and Bernstein, R.B. (1974). Energy disposal and energy consumption in elementary chemical reactions: the information theoretic approach. Acc. Chem. Research *7*, 393 - 400.

Levine, R.D. and Bernstein R.B. (1976) in *Dynamics of Molecular Collisions*. (Edited by W.H. Miller). Part B., p. 323. New York.

Maroulis, G., Sana, M., and Leroy, G. (1981). Molecular properties and basis set quality: an approach based on Information Theory. Int. J. Quantum Chem. *19*, 43 - 60.

Martin, Y.C. (1978). *Quantitative Drug Design*. Decker, New York.

Mathai, A.M., and Rathie, P.N. (1975). *Basic Concepts in Information Theory and Statistics, Axiomatic Foundations and Applications*. Wiley, New York.

Mc Eliece, R.J. (1977). *The Theory of Information and Coding*. Addison-Wesley, Reading, Massachusetts.

Mc Weeny, R., and Sutcliffe, B.T. (1969). *Methods of Molecular Quantum Mechanics*, Academic, London.

Mekenyan, O., Bonchev, D., and Trinajstić, N. (1979 a). Topological rules for spirocompounds. Math. Chem. (MATCH) *6*, 93 - 115.

Mekenyan, O., Bonchev, D., and Trinajstić, N. (1980). Modeling the thermodynamic properties of molecules. Int. J. Quantum Chem. Symp. *18*, 369 - 380.

Mekenyan, O., Bonchev, D., and Trinajstić, N. (1981 a). On algebraic characterization of bridged polycyclic compounds. Int. J. Quantum Chem. *19*, 929 - 955.

Mekenyan, O., Bonchev, D., and Trinajstić, N. (1981 b). A topological characterization of cyclic structures with acyclic branches. Math. Chem. (MATCH) *11*, 145 - 168.

Mekenyan, O., Bonchev, D. and Trinajstić, N. (1982). Molecular complexity and properties of cyclic molecules with acyclic branches. J. Comput. Chem. (submitted).

Merrifield, R.E. and Simmons, H.E. (1981 a). Enumeration
of structure-sensitive graphical subsets: Theory. Proc.
Natl. Acad. Sci. USA, *78*, 692 - 695; idem. (1981 b).
Enumeration of structure-sensitive graphical subsets:
Calculations. Proc. Natl. Acad. Sci. USA *78*, 1329 - 1332.

Mizutani, K., Kawasaki, K., and Hosoya, H. (1971). Tables
of non-adjacent numbers, characteristic polynomials, and
topological indices. I. Tree graphs. Nat. Sci. Rept. Ochanomizu Univ. (Jpn) *22*, 39 - 58.

Morgan, H.L. (1965). The generation of a unique machine
description for chemical structures. A technique developed at Chemical Abstract Service. J. Chem. Doc. *5*,
107 - 113.

Morovitz, H. (1955). Some order- disorder considerations
in living systems. Bull. Math. Biophys. *17*, 81 - 86.

Mowshowitz, A. (1968 a). Entropy and the complexity of
graphs: I. An index of the relative complexity of a
graph. Bull. Math. Biophys. *30*, 175 - 204.

Mowshowitz, A. (1968 b). Entropy and the complexity of
graphs: II. The information content of digraphs and infinite graphs. Bull. Math. Biophys. *30*, 225 - 240.

Mowshowitz, A. (1968 c). Entropy and the complexity of
graphs: III. Graphs with prescribed information content.
Bull. Math. Biophys. *30*, 387 - 414.

Mowshowitz, A. (1968 d). Entropy and the complexity of
graphs: IV. Entropy measures and graphical structure.
Bull. Math. Biophys. *30*, 533 - 546.

Papazova, D., Dimov, N., and Bonchev, D. (1980). Calculation of gas chromatographic retention indices of isoalkanes based on a topological approach. J. Chromatogr.
188, 297 - 303.

Peev, T., Bonchev, D., Rousseva, B., and Dimitrov, A.
(1972). Information Study of 2β - stable atomic nuclei.
God. Vissh Khim.- Tekhnol. Inst., Burgas, Bulg. *9*,
73 - 86 (in Bulgarian).

Peev, T., Bonchev, D., Dimitrov, A., and Rousseva, B. (1974). Differential information characteristics of the nucleon distribution in 2β - stable isotopes. God. Vissh. Techn. Uch. Zav. Fiz., Sofia, Bulg. *11*, 87 - 98 (in Bulgarian).

Petrov, T.G. (1970 a). in *Trudi Symposiuma 2nd 1969*. (Edited by L.A. Aleksandrov) p. 61, Sib. Otd. Akad. Nauk SSSR, Novosibirsk (in Russian).

Petrov, T.G. (1970 b). On the measure of complexity of geochemical systems from the viewpoint of information theory. DAN SSSR, *191*, 924 - 926 (in Russian).

Petrov, T.G. (1971). Providing the basis of a new version of the general classification of geochemical systems. Vest. Leningrad. Univ. Geol., Geogr. *3*, 30 - 38 (in Russian).

Platt, J.R. (1947). Influence of neighbour bonds on additive bond properties in paraffins. J. Chem. Phys. *15*, 419 - 420.

Platt, J.R. (1952). Prediction of isomeric differences in paraffin properties. J. Phys. Chem. *56*, 328 - 336.

Polansky, O.E. (1980). Unpublished.

Polansky, O.E. (1975). Polya's method for the enumeration of isomers. Math. Chem. (MATCH) *1*, 11 - 31.

Purcell, W.P., Bass, G.E., and Clayton, J.M. (1973). *Strategy of Drug Design*. Wiley, New York.

Quastler, H., Editor (1953). *Essays on the Use of Information Theory in Biology*. University of Illinois, Urbana.

Rackow, B. (1963). Group-theoretic reasonings to introduce the entity of molbit in molecular-biological physical chemistry. Z. Chem. *3*, 268 (in German).

Rackow, B. (1967 a). Aufprägung von molecülare information auf oberflächen von organischen hochpolymeren. Z. Chem. *7*, 398 - 400.

Rackow, B. (1967 b). Ableitung der "Grundgleichung der informationschemie" fur molekulargeprägte polymere. Z. Chem. *7*, 472 - 474.

Rackow, B. (1969). Moleculare information eines anorganischen systems ionengitter-oberfläche/adsorbierte metall-ionen. Z. Chem. *9*, 318 - 319.

Randić, M. (1975). On characterization of molecular branching. J. Am. Chem. Soc. *97*, 6609 - 6615.

Randić, M. (1979). Characterization of atoms, molecules and classes of molecules based on path enumerations. Math. Chem. (MATCH) *7*, 5 - 64.

Randić, M., Brissey, G.M., Spencer, R.B., and Wilkins, C.L. (1979). Search for all self-avoiding paths for molecular graphs. Comp. Chem. *3*, 5 - 13.

Rashevsky, N. (1955). Life, information theory, and topology. Bull. Math. Biophys. *17*, 229 - 235.

Rashevsky, N. (1960). Life, information theory, probability, and physics. Bull. Math. Biophys. *22*, 351 - 364.

Ray, S.K., Basak, S.G., Raychaudhury, C., Roy, A.B., and Ghosh, J.J. (1981 a). Quantitative structure-activity relationship studies of bioactive molecules using structural information indices. Ind. J. Chem. *20B*, 894 - 897.

Ray, S.K., Basak, S.C., Raychaudhury, C., and Ghosh, J.J. (1981 b), in Abstracts, Proceedings of the Annual Convention of Chemists, Indian Chem. Society.

Ray, S.K., Basak, S.C., Raychaudhury, C., and Ghosh, J.J. (1981 c), in Abstracts, Intern. Conf. on Communication Circuits and Systems, Calcutta.

Ray, S.K., Basak, S.C., Raychaudhury, C., Roy, A.B., and Ghosh, J.J. (1982 b). A quantitative structure-activity relationship (QSAR) study of triazinones, alkyl carbamates, and β-piridylalkylamines using information-theoretic topological indices. J. Comp. Chem. (in press).

Ray, S.K., Basak, S.C., Raychaudhury, C., Roy, A.B. and Ghosh, J.J. (1982 a). A quantitative structure-activity relationship (QSAR) study of N-alkylnorketobemidones and triazinones using structural information content. Arzneim-Forsch. *32*, 322 - 325.

Raychaudhury, C., Basak, S.C., Roy, A.B., and Ghosh, J.J. (1980). Quantitative structure-activity relationship (QSAR) studies of pharmacological agents using topological information content. Indian Drugs *18*, 97 - 102.

Raychaudhury C., Basak, S.C., Ray, S.K., Roy, A.B., and Ghosh, J.J. (1982). Graph-theoretical invariant and thermodynamic property: A QSAR study of alcohols, 19th Annual Meeting, Society of Eng. Sci. Inc. University of Missouri Rolla, Rolla.

Read, R.C. and Milner, R.S. (1978). *A New System for the Designation of Chemical Compounds for the Purposes of Data Retrieval. I. Acyclic Compounds.* Res. Rep. CORR 78 - 42, University of Waterloo, Ontario.

Read, R.C. (1980). *II. Cyclic Compounds.* Res. Rep. CORR 80 - 7, University of Waterloo, Ontario.

Renyi, A. (1965). On the foundations of Information Theory. Rev. Intern. Statist. *3*, 1 - 14.

Rousseva, B., Dimitrov, A., and Peev, T. (1976). Mean information content of the nucleons in the atomic nuclei of 2β-stable isotopes. God. Vissh.Khim.-Tekhnol. Inst., Burgas, Bulg. *11*, 351 - 356 (in Bulgarian).

Rousseva, B., and Bonchev, D., (1978). Information-theoretic variant of the systematic of nuclides. Math. Chem. (MATCH) *4*, 173 - 192.

Rousseva, B. and Bonchev, D. (1979). Empirical correlations between the nuclear binding energy and electron energy of isodifferent nuclides. Radiochem. Radioanal. Lett. *40*, 41 - 50.

Rousseva, B. and Bonchev, D. (1980) Prediction of the nuclear binding energies of the nuclides of period VII. Radiochem. Radioanal. Lett. *45*, 341 - 346.

Rouvray, D.H. (1971). Graph theory in Chemistry. R.I. C. Rev. *4*, 173 - 195.

Rouvray, D.H. (1973). The search for useful topological indices in chemistry. Amer. Sci. *61*, 729 - 735.

Rouvray, D.H. (1974). Uses of graph theory. Chem. Brit. *10*, 11 - 15.

Rouvray, D.H. (1975 a). The chemical applications of Graph Theory. Math. Chem. (MATCH) *1*, 61 - 70.

Rouvray, D.H. (1975 b). The value of topological indices in chemistry. Math. Chem (MATCH) *1*, 125 - 134.

Rouvray, D.H. and Crawford, B.C. (1976). The dependence of physicochemical properties on topological factors. S.Afr. J. Sci. *72*, 47 - 61.

Rouvray, D.H. and Balaban, A.T. (1979), in *Applications of Graph Theory*. (Edited by R.J. Wilson and L.W. Beineke). p. 177, Academic, London.

Roy, A.B. Raychaudhury, C., Ray, S.K., Basak, S.C., and Ghosh, J.J. (1982). Information theoretic topological indices of a molecular graph and their application in QSAR (Work in preparation).

Sablić, A. and Trinajstić, N. (1981). QSAR. The role of topological indices. Acta Pharm. Jug. *31*, 189 - 214.

Sakamoto, K., Yee, W.T., and I'Haya, Y.J. (1977). Information theoretical aspects of molecular properties. Part 2. The application to hydrocarbons (chain and cyclic paraffins and monoolefins). Rep. Univ. Electro-Comm. *27*, 227 - 244.

Sarkar, R., Roy, A.B. and Sarkar, P.K. (1978). Topological

information content of genetic molecules. I. Math. Biosci. *39*, 299 - 312.

Schroedinger, E. (1944). *What is Life?* University Press, Cambridge, England.

Scillard, L. (1929). Ueber die Entropieverminderung in einem thermodinamischen System bei Eingriffen intelligenter Wesen. Z. Phys. *53*, 840 - 856.

Sears, S.B., Parr, R.G., and Dinur, U. (1980). On the quantummechanical kinetic energy as a measure of the information in a distribution. Isr. J. Chem. *19*, 165 - 173.

Seybold, P.G. (1976). Why are there four bases in DNA? Int. J. Quantum Chem., Quantum Biology Symp. No. 3, 39 - 43.

Shannon, C., and Weaver, W. (1949). *Mathematical Theory of Communication*. University of Illinois, Urbana.

Smolenski, E.A. (1964). Graph theory application to the calculations of structural - additive properties of hydrocarbons. Zh. Fiz. Khim. *38*, 1288 - 1291.

Tatevskii, V.M. (1953). *Khimicheskoe Stroenye Uglevodorodov i Zakonomernosti v Ikh Fiziko-Khimichestikh Svoistv*. Nauka, Moscow.

Trinajstić, N. (1982). *The Chemical Graph Theory*. CRC, New York (in press).

Trinajstić, N., Protić, G., Svob, V., and Deur-Siftar, Dj. (1979). Calculation of retention indexes in gas chromatography by means of structural parameters. Alkylbenzenes. Kem. Ind. (Zagreb) *28*, 527 - 535.

Trincher, K.S. (1964). *Biology and Information*. Nauka, Moscow (in Russian).

Trucco, E. (1956 a). A note on the information content of graphs, Bull. Math. Biophys. *18*, 129 - 135.

Trucco, E. (1956 b). On the infromation content of graphs: Compound Symbols; Different states for each point. Bull. Math. Biophys. *18*, 237 - 253.

Valentinuzzi, M. and Valentinuzzi, M.E. (1962). Contribution to the study of the information content of chemical structures (in Spanish). Anal. Soc. Cientif. Ar-
gentina 1 - 86.

Valentinuzzi, M. and Valentinuzzi, M.E. (1963). Information content of chemical structures and some possible
biological applications. Bull. Math. Biophys. *25*, 11 - 27

Wiener, H. (1947). Structural Determination of Paraffin
Boiling Points. J.Am. Chem. Soc. *69*, 17 - 20.

Wiener, N. (1948). *Cybernetics.*, Wiley, New York.

Wigner, E. (1931). *Gruppentheorie und ihre Anwendungen
auf die Quantenmechanik der Atomspektren.* Friedr. Vie-
weg, Braunschweig.

Wilson, R.J. (1972). *Introduction to Graph Theory.* Oliv-
er and Boyd, Edinburgh.

Vistelius, A.B. (1964). Information characteristic of
frequency distributions in geochemistry. Nature, *202*,
1206.

Woodward, R.B., and Hoffman, R. (1970). *The Conservation of Orbital Symmetry.* Chemie, Weinheim.

Ursul, A.D. (1966). An information criterion of evolution in nature. Filosofskie Nauki No. 2, 43 - 51 (in
Russian).

Yatsimirsky, K.B. (1975). Applications of graph theory
in chemistry. Int. Chem. Eng. *15*, 7 - 20.

Yee, W.T., Sakamoto, K., and I'Haya, Y.J. (1976). Information theoretical aspects of molecular properties.
Part 1. A theoretical study of information content of
organic molecules. Rep. Univ. Electro-Comm. *27*, 53 - 63.

Zhdanov, Y.A., and Minkin, V.I. (1966). *Correlation
Analysis in Organic Chemistry.* University of Rostov,
Rostov, USSR (in Russian).

Zhdanov, Y.A. (1967). On the amount of information in
the molecules of bioorganic compounds. Biofiz. *12*, 715 -
717 (in Russian).

Zhdanov, Y.A. (1974). A correlation in genetic code.
DAN SSSR, *217*, 456 - 459 (in Russian).
Zhdanov, Y.A. (1979). *Information Entropy in Organic
Chemistry*. Rostov University, Rostov, USSR (in Russian).
Zemanek, H. (1959). *Elementare Informationstheorie*. Ol-
denbourg, Wien.

Glossary

A - mass number
A(G) - total vertex adjacency
$A(G)$ - vertex adjacency matrix
$A_E(G)$ - total edge adjacency
$A_E(G)$ - edge adjacency matrix
α - electronic state
B(G) - the Balaban centric index
β - isodifferent number
C - cycle
$c_i(V)$ - vertex cyclic degree
$c_i(E)$ - edge cyclic degree
$C_E(G)$ - total edge cyclicity
$C_V(G)$ - total vertex cyclicity
$C(G)$ - (edge) cycle matrix
$C_V(G)$ - vertex cycle matrix
$C'_{n,r}$ - combinations of n elements of class r taken r at
          a time with repetitions
d_{ij} - distance between vertices (edges) i and j
d_i - (vertex) distance degree
$d_i(E)$ - edge distance degree
$D(G)$ - (vertex) distance matrix
$D_E(G)$ - edge distance matrix
DG - directed graph
E - edge
E_b - nuclear binding energy
E_a - electronic energy
E - edge chromatic decomposition
e_i - edge degree
ε - electropy
∈ - (in set theory) belongs to
G - graph
H_i - entropy of the outcome i

$H(P)$ - mean entropy of the probability distribution P

H_{melt} - heat of melting

H_{subl} - heat of sublimation

ΔH^o_{298} (form) - standard heat of formation

ΔH^o_{298} (gas, comb) - standard heat of combustion in the gas state

ΔH^o_{298} (liq, comb) - standard heat of combustion in the liquid state

$x_R(G)$ - the Randić connectivity index

I_1, I_2 - first and second ionization potential

I - quantity of information (information content, information index)

I - total information index (per structure)

$\bar{I}$ - mean information index (per structural element)

ΔI - differential information index

$^{v(E)}I_x^{E(M)}(G)$ - information index on the distribution of graph vertices v or edges (E) according to the equivalence E or magnitude (M) of the graph-invariants x, respectively

$\tilde{I}$ - information aromaticity index

i - mean isomer degeneracy

$I(G)$ - incidence matrix

$J(G)$ - the Balaban distance connectivity index

ξ - the Slater orbital exponent

$k(G)$ - (vertex) chromatic number

$k_E(G)$ - edge chromatic number

lb - logarithm at basis two

ΔL_{298} - standard heat of vaporization

MG - multigraph

$M(G)$ - the Zagreb index

N - the total number of elements in the structure (the cardinality of the set of elements)

$N!$ - permutations of N elements

$\{N_1, N_2, \ldots, N_k\}$ - unordered set of structure elements distributed into k equivalence classes

$(N_1, N_2, \ldots, N_k)$ - ordered set

N_i - cardinality of the subset i

$[N/2]$ - the Gauss square brackets

P - parachor

p_i - probability of the outcome i

$P(p_1, p_2, \ldots, p_k)$ - probability distribution

$P(G,k)$ - non-adjacent number of a graph

$v(E)_{I_x} E(M)$ - vertex v (or edge E) partition according to the equivalence E or magnitude (M) of the graph-invariants x

QSAR - quantitative structure-activity relationships

R_{cov} - covalent radius

r_i - vertex radius

$r_i(E)$ - edge radius

RI - gas chromatographic retention index

ρ - steric parameter

S^o_{gas}, S^o_{solid} - standard entropy in the gas and solid state

σ^* - the Taft polar constant

T_B, T_M - boiling and melting points

$\cup$ - union (in set theory)

V - vertex

V_A - atomic volume

v_i - vertex degree

$W(G)$ - the Wiener index

ω - mean oxidation number

z - atomic number

$Z(G)$ - the Hosoya index

Index